Marco Cavallari

Antigen-presentation of non-peptidic antigens

Marco Cavallari

Antigen-presentation of non-peptidic antigens

Trafficking, loading and presentation of lipids and small metabolites

Südwestdeutscher Verlag für Hochschulschriften

Impressum/Imprint (nur für Deutschland/only for Germany)
Bibliografische Information der Deutschen Nationalbibliothek: Die Deutsche Nationalbibliothek verzeichnet diese Publikation in der Deutschen Nationalbibliografie; detaillierte bibliografische Daten sind im Internet über http://dnb.d-nb.de abrufbar.

Alle in diesem Buch genannten Marken und Produktnamen unterliegen warenzeichen-, marken- oder patentrechtlichem Schutz bzw. sind Warenzeichen oder eingetragene Warenzeichen der jeweiligen Inhaber. Die Wiedergabe von Marken, Produktnamen, Gebrauchsnamen, Handelsnamen, Warenbezeichnungen u.s.w. in diesem Werk berechtigt auch ohne besondere Kennzeichnung nicht zu der Annahme, dass solche Namen im Sinne der Warenzeichen- und Markenschutzgesetzgebung als frei zu betrachten wären und daher von jedermann benutzt werden dürften.

Verlag: Südwestdeutscher Verlag für Hochschulschriften GmbH & Co. KG
Heinrich-Böcking-Str. 6-8, 66121 Saarbrücken, Deutschland
Telefon +49 681 37 20 271-1, Telefax +49 681 37 20 271-0
Email: info@svh-verlag.de

Approved by: Basel, Universität Basel, Inauguraldissertation, 2010

Herstellung in Deutschland:
Schaltungsdienst Lange o.H.G., Berlin
Books on Demand GmbH, Norderstedt
Reha GmbH, Saarbrücken
Amazon Distribution GmbH, Leipzig
ISBN: 978-3-8381-1432-3

Imprint (only for USA, GB)
Bibliographic information published by the Deutsche Nationalbibliothek: The Deutsche Nationalbibliothek lists this publication in the Deutsche Nationalbibliografie; detailed bibliographic data are available in the Internet at http://dnb.d-nb.de.

Any brand names and product names mentioned in this book are subject to trademark, brand or patent protection and are trademarks or registered trademarks of their respective holders. The use of brand names, product names, common names, trade names, product descriptions etc. even without a particular marking in this works is in no way to be construed to mean that such names may be regarded as unrestricted in respect of trademark and brand protection legislation and could thus be used by anyone.

Publisher: Südwestdeutscher Verlag für Hochschulschriften GmbH & Co. KG
Heinrich-Böcking-Str. 6-8, 66121 Saarbrücken, Germany
Phone +49 681 37 20 271-1, Fax +49 681 37 20 271-0
Email: info@svh-verlag.de

Printed in the U.S.A.
Printed in the U.K. by (see last page)
ISBN: 978-3-8381-1432-3

Copyright © 2012 by the author and Südwestdeutscher Verlag für Hochschulschriften GmbH & Co. KG and licensors
All rights reserved. Saarbrücken 2012

Fama refert nostros te, Fidentine, libellos
Non aliter populo quam recitare tuos.

Si mea vis dici, gratis tibi carmina mittam:
Si dici tua vis, hoc eme, ne mea sint.

Marcus Valerius Martialis
Epigrammaton liber I, XXIX

I will always be grateful to all the people that helped me to achieve the goals of so many projects with their enthusiasm and tremendous efforts; especially the members of the Experimental Immunology laboratory, past and present, for their unconditional help, insightful discussions, critical but constructive input on the work in general and on the thesis work in particular, and for their friendship.

This work was supported by the Swiss National Science Foundation (grant 3100AO-109918), by EEC grant MOLSTROKE (Molecular basis of vascular events leading to thrombotic stroke, LSHM-CT-2004 Contract Number 005206) and I was awarded a doctoral fellowship by the Roche Research Foundation (grant 2006-47).

Further I would like to thank the "Reisefonds für den akademischen Nachwuchs der Universität Basel" for the subsidies to visit the 4th International NKT Cell & CD1 Workshop [De Libero et al. 2007 b] held October 4 to 8 2006 at the Abbazia di Spineto, in the heart of Tuscany.

Impressum

Typeset by TeXnicCenter and MiKTeX.

Published under the copyright standard (for details see appendices, p.207) ...
... 'Creative Commons' 'Attribution No Derivatives'.

Summary

T cells recognize a broad variety of antigens, including peptides, lipids and non-peptidic phosphorylated metabolites. Clarification of the rules rendering non-peptidic molecules immunogenic is essential to understand and to influence the reactions of the immune system to this class of substances in health and disease. Despite recent advances in research about immune responses to non-peptidic compounds, important issues remain unanswered. Molecular mechanisms governing the immunogenicity of non-peptidic ligands such as their cell internalization, trafficking within intracellular organelles, association with dedicated antigen-presenting molecules, induction of central and peripheral tolerance, and finally their role in autoimmune diseases as well as in protection during infections are unknown to date.

The aims of this thesis were to assess some of the immunological functions and cell biological rules governing the immunogenicity of non-peptidic antigens, with particular emphasis on cell trafficking of non-peptidic antigens and antigen-presenting molecules. It focused on **(i)** the antigen reactivity and presence of human invariant natural killer T (iNKT) cells in diseases, **(ii)** the role of CD1a trafficking in lipid antigen presentation by this protein, and **(iii)** the requirements of membrane translocation of phosphorylated mevalonate metabolites that stimulate human T cell receptor (TCR) $\gamma\delta$ cells.

With the development of alpha-galactosylceramide (αGC)-loaded soluble CD1d dimers, which specifically interact with the TCR of iNKT cells, we have the perfect tool in our hands to perform detailed studies on iNKT cells. Analysis of the iNKT cells in blood unveiled large differences in their fluorescence intensity suggesting the presence of semi-invariant iNKT TCR with large disparities in the affinity for the αGC-CD1d complex. Unexpectedly, established iNKT cell clones showed no correlation between CD1d dimer-staining levels and αGC reactivity, indicating that additional mechanisms control responsiveness of iNKT cells, at least to this lipid antigen.

The identification of lipid antigens stimulating exclusively some desired functions in human iNKT cells might lead to new medical therapies or vaccines. To screen a variety of synthetic lipids for their capacity to activate iNKT cells, we devised an *in vitro* model based on plastic-bound CD1d. Piperidinones, molecules with a ceramide- or sphingosine-like structure, revealed that a single lipid tail is sufficient to form stimulatory complexes with CD1d. Interestingly, piperidinones preferentially induce T_H1-like cytokines, predicting a possible role as novel leader molecules to functionally direct iNKT cell responses deployable in clinical therapies.

The balance of proinflammatory T_H1 to regulatory T_H2 cytokines is well-known to be decisive for the outcome of many diseases. Atherosclerosis (ATH) is a chronic inflammatory disease characterized by lipid accumulation in plaques. The disease is complicated by cardiovascular events provoked by plaque rupture or erosion. Because inflammation participates in lesion progression and rupture of plaques, the identification of its causes and of the culprit leukocyte populations involved in plaque destabilization is crucial for effective prevention of cardiovascular events. We used CD1d dimers to detect and characterize iNKT cells in ATH patients. We found that, in human atherosclerotic lesions, the abundance of $CD1d^+$ antigen-presenting cells (APC) and of iNKT cells correlates with disease severity and activity. $CD1d^+$ cells colonize advanced plaques in symptomatic patients and are most abundant in plaques with concomitant signs of ectopic neovascularization. In plaques, the frequency of iNKT cells among total T cells exceeds the one in blood. After having successfully isolated iNKT cell lines from plaque tissue, we showed that they promptly release proinflammatory cytokines upon lipid antigen stimulation and promote endothelial cell migration and microvascular sprout formation *in vitro*. This functional proangiogenic activity is ascribed to interleukin-8 released by iNKT cells after lipid recognition. These findings introduce iNKT cells as novel candidates to induce plaque neovascularization and destabilization in human ATH. Targeting iNKT cells could lead to late stage ATH treatment.

Another approach to understand the role of lipid-specific immune responses is to investigate the molecular rules of lipid-CD1 complex formation. Lipids distribute, due to their physicochemical properties or with the help of specific transporters and lipid transfer proteins, to different intracellular compartments and membrane domains. Thus, it is advantageous for the immune system to utilize multiple CD1 isoforms, each with a distinct trafficking pattern, to facilitate sampling of lipid antigens localized in various membranes. Several studies have addressed trafficking of CD1 isoforms. However, the molecular mechanisms are known in only a few cases.

We identified invariant chain (Ii) and lipid rafts as key regulators of CD1a organization on the surface of APC and of its immunological function as antigen-presenting molecule. Colocalization of CD1a with Ii is dependent on raft integrity and CD1a internalization is increased by Ii. The localization of CD1a in lipid rafts is functionally relevant as raft disruption inhibits CD1a-restricted antigen presentation.

Moreover, we found that CD1a is internalized independently of clathrin and dynamin and that it follows a Rab22a- and adenosine diphosphate ribosylation factor (ARF) 6-dependent recycling pathway, similar to other clathrin-independent cargo. Posttranslational S-acylation of the CD1a cytoplasmic tail may occur but neither determines the rate of internalization nor recycling nor its localization to detergent-resistant membrane microdomains. These findings place CD1a close to major histocompatibility complex (MHC) class I in its trafficking routes although CD1a loads lipids in recycling endosomes and not in the endoplasmic reticulum as MHC class I.

Strikingly, the glycolipid antigen sulfatide was found localized predominantly to early and recycling endosomes where CD1a is located. Swapping the cytoplasmic tail of CD1a for the one of CD1b and hence targeting the CD1a protein to the late endosomal and lysosomal compartments decreases its capacity to present sulfatide and shortens the half-life of stimulatory complexes. Thus, the physiological intracellular trafficking route of CD1a is critical for efficient presentation of lipid antigens that traffic through the early endocytic and recycling pathways.

Intracellular trafficking of another class of non-peptidic antigens, namely the phosphorylated metabolites which stimulate human TCR $\gamma\delta$ cells expressing the Vγ9/Vδ2 heterodimer, was examined. These T cells recognize a family of structurally related compounds produced in the eukaryotic mevalonate and prokaryotic methylerythritol phosphate (MEP) pathways. The endogenous self-ligands are generated within the cytoplasm and must cross the membrane in order to associate with dedicated antigen-presenting molecules, which remain unknown at present.

Using an *in vitro* transport assay, we demonstrated that the multidrug resistance-associated protein (MRP) 5 transporter is involved in membrane translocation of antigenic phosphorylated metabolites. Confocal microscopy illustrated that MRP5 is located in membranes of both endoplasmic reticulum and early endosomes. Both the intracellular localization and active role in antigen transport confer an immunological function to MRP5, resembling that of TAP (transporter associated with antigen processing) transporters involved in peptide antigen translocation. This indicates a similar strategy used for antigen presentation to TCR $\alpha\beta$ and $\gamma\delta$ T cells.

In conclusion, these studies have underlined the physiological relevance of T cells recognizing non-peptidic ligands and have revealed unanticipated molecular mechanisms controlling the efficient presentation of such antigens.

Contents

Summary		iii
A Short History of Non-classical Immunology		1
I.	Introduction	5
1.	The immune system	7
2.	**Non-peptidic molecules**	**9**
2.1.	Lipid evolution and repertoires	9
2.2.	Lipid localization and metabolism	9
2.3.	Lipid cellular trafficking	12
2.4.	Phosphorylated metabolites	13
2.5.	Foreign and self non-peptidic antigens	13
3.	**Antigen-presenting molecules (APM)**	**15**
3.1.	CD1	15
3.1.1.	CD1 evolution and morphology	15
3.1.2.	CD1 tissue expression and cell distribution	18
3.1.3.	CD1 assembly, trafficking and loading	19
3.2.	APM for IPP and gamma-delta T cells	21
4.	**Non-peptidic antigen-specific T cells**	**23**
4.1.	alpha-beta T cells	23
4.1.1.	iNKT cells	23
4.1.2.	NKT cells	25
4.2.	gamma-delta T cells	26
II.	Materials and Methods	29
5.	**Materials**	**31**
5.1.	Antibodies	31
5.2.	Cell lines and clones	34
5.3.	Compounds	36
5.4.	Cytokines and chemokines	37
5.5.	DIMER	37
5.6.	Media and buffers	39
5.6.1.	Media basis	39
5.6.2.	Buffers	39

5.7.	Primers	40
5.7.1.	Mycoplasma	40

6. Cell culture techniques — 41
6.1.	Cloning, cells	41
6.2.	Cloning, lines	41
6.3.	Cytospin	41
6.4.	Depletion and enrichment	42
6.4.1.	Isolation of lymphocytes from tissue	42
6.4.2.	Isolation of peripheral blood mononuclear cells (PBMC)	42
6.5.	Fixation	43
6.6.	Maintenance	43
6.6.1.	Freezing	43
6.6.2.	Restimulation	44
6.6.3.	Thawing	44
6.7.	Sorting	44

7. Immunological techniques — 45
7.1.	Antigen presentation assay	45
7.1.1.	Antigen presentation assay, chase	45
7.1.2.	Antigen presentation assay, competition	45
7.1.3.	Antigen presentation assay, pulse	45
7.1.4.	Antigen presentation assay, standard	45
7.2.	Cell tracing and tracking	45
7.3.	Co-culture assay	46
7.4.	Cytotoxicity assay	46
7.5.	Immunoprecipitation	46
7.6.	Confocal laser scanning microscopy (CLSM)	46
7.7.	Enzyme-linked immunosorbent assay (ELISA)	46
7.8.	Fluorescence-activated cell sorting (FACS)	47
7.9.	Fluorescence microscopy	47
7.10.	Proliferation assay	47
7.10.1.	Thymidine uptake	47
7.10.2.	Carboxyfluorescein succinimidyl ester (CFSE)	47

8. Molecular techniques — 49
8.1.	Genomic deoxyribonucleic acid (gDNA) preparation	49
8.2.	Electrophoresis	49
8.2.1.	Agarose gel electrophoresis	49
8.3.	Polymerase chain reaction (PCR)	49
8.4.	Vesicle uptake studies	49

9. Patients and healthy donors — 51
9.1.	Atherosclerosis patient cohort from Cantonal Hospital Luzern	51
9.2.	Atherosclerosis patients from Cantonal Hospital Bruderholz	53
9.3.	Healthy donor cohort from Blutspende beider Basel	53

10. Software and statistical analyses — 55
10.1.	CLSM software	55
10.1.1.	ImageJ	55

10.1.2.	Imaris	55
10.1.3.	ZEN	55
10.2.	FACS software	55
10.2.1.	FCS Express	55
10.2.2.	FlowJo	55
10.2.3.	Summit	56
10.3.	Fluorescence microscopy software	56
10.3.1.	cell-P	56
10.4.	Graphic software	56
10.4.1.	GIMP	56
10.5.	Line art software	56
10.5.1.	GraphPad Prism	56
10.6.	Statistical software	56
10.6.1.	GraphPad Prism	56
10.6.2.	R	56

III. Results and Discussion — 57

11. CD1-DIMER, a T cell envision tool — 59
- 11.1. Clonal brightness — 63
- 11.2. A new analogue of alpha-galactosylceramide — 71
- 11.3. Sphingolipid analogues based on 7-oxasphingosine and 7-oxaceramide — 73
- 11.4. 7-aza- and 7-thiasphingosines — 76
- 11.5. 4,5,6-tri-substituted piperidinones as conformationally restricted ceramide analogues — 79
- 11.6. Final conclusions and outlook — 84

12. iNKT cells link inflammation and neovascularization in human ATH — 87
- 12.1. CD1d-expressing cells in atherosclerotic lesions are a sign of arterial vulnerability — 90
- 12.2. iNKT cells are found in atherosclerotic lesions — 90
- 12.3. Reduction of circulating iNKT cells in ATH patients — 92
- 12.4. Characterization of proatherosclerotic activity of iNKT cells — 97
- 12.5. Soluble factors released by iNKT cells promote EC migration — 97
- 12.6. IL-8 is produced by iNKT cells and induces EC migration — 97
- 12.7. Discussion and Outlook — 100

13. Rafting and docking with CD1a — 105
- 13.1. CD1a associates with Ii, HLA-DR and CD9 — 107
- 13.2. Ii silencing increases CD1a at the cell surface — 108
- 13.3. CD1a partition to detergent-resistant membrane microdomains is necessary for efficient exogenous antigen presentation — 110
- 13.4. CD1a chimeras – tail to traffic twists — 111
- 13.5. Tail-deletion does not change the subcellular distribution of CD1a — 114
- 13.6. CD1a TD shows normal surface expression, internalization and recycling — 115
- 13.7. Tail-deletion does not affect presentation of sulfatide to T cells — 117
- 13.8. CD1a surface expression is not changed by inhibition of the clathrin pathway — 118
- 13.9. CD1a follows a Rab22a-dependent recycling pathway — 118
- 13.10. CD1a is internalized and recycled by an adenosine diphosphate ribosylation factor 6-dependent pathway — 119
- 13.11. CD1b does not accumulate in ARF6-Q67L-positive enlarged vesicles — 121

13.12.	Relocation of CD1a to lysosomes by providing residues of CD1b cytoplasmic tail	121
13.13.	Colocalization of CD1a and sulfatide	125
13.14.	WT CD1a transfectants present sulfatide more efficiently than CD1aab chimera	127
13.15.	CD1a:CD1b chimeras have shorter-lived ability to stimulate T cells	127
13.16.	Conclusions and outlook	128

14. MRP5 transporter is required for TCR gamma-delta-ligand transport — 135

14.1.	The TCR gamma-delta-ligand transporter belongs to the ABC family	138
14.2.	Involvement of ABCC subfamily in TCR gamma-delta-ligand transport	140
14.3.	MRP5 overexpression increases TCR gamma-delta-cell stimulation	141
14.4.	MRP5 silencing affects stimulation of TCR gamma-delta cells	141
14.5.	MRP5 transports IPP across membranes	142
14.6.	MRP5 resides in ER and EE membranes	145
14.7.	Conclusions and outlook	148

IV. General Conclusions and Future Work — 151

Appendices — 155

Abbreviations — 157

Index — 161

Bibliography — 165

SOP — 195

1.	SOP PBMC	195
2.	SOP mycoplasma	199

Documents — 207

3.	Creative Commons License	207

List of Figures and Tables

List of Figures

2.1.	Spatial lipid organization to form membrane asymmetry and domains	11
3.1.	Evolutionary hypothesis of the human CD1 multigene family	17
11.1.	Alpha-galactosylceramide-loaded human CD1d DIMER	62
11.2.	Establishing the DIMER system	63
11.3.	Sorting of dull and bright DIMER-positive iNKT cells from PBMC	65
11.4.	Plating efficiency of the BG clones	66
11.5.	Clonal brightness	66
11.6.	Correlation of the TCR alpha- and beta-chain with the DIMER	67
11.7.	Relation of the classical CD4 and CD8 co-receptors to the DIMER	67
11.8.	Binding to CD1d and failure to activate iNKT cells	72
11.9.	Competition of triazoles with alpha-galactosylceramide	75
11.10.	A triazole with sensitivity to low pH	75
11.11.	Competition of azasphingosines with alpha-galactosylceramide on plastic CD1d	77
11.12.	Competition of azasphingosines with alpha-galactosylceramide on APC	78
11.13.	Single tail piperidinones presented by living APC activate iNKT cells	83
11.14.	Single tail piperidinones presented by plate-bound CD1d activate iNKT cells	84
11.15.	Double tail piperidinones presented by living APC activate iNKT cells	85
11.16.	Double tail piperidinones presented by plate-bound CD1d activate iNKT cells	86
12.1.	APC in atherosclerotic lesions	91
12.2.	Specific identification of iNKT cells in ATH lesions	93
12.3.	Polarization of the TCR on iNKT cells towards CD1d on APC in ATH lesions	94
12.4.	iNKT cells from atherosclerotic plaque tissue	95
12.5.	Half maximal effective concentration of alpha-galactosylceramide with iNKT cells from atherosclerotic plaque tissue	96
12.6.	Circulating iNKT cells are reduced in ATH patients	96
12.7.	Antigen activation of iNKT cells increases sprout outgrowth from EC spheroids	98
12.8.	Antigen activation of iNKT cells promotes EC migration	99
12.9.	Antigen activated iNKT cells produce IL-8 which promotes EC migration	101
13.1.	CD1a associates with CD9, HLA-DR and Ii	109
13.2.	Ii interference induces CD1a accumulation at the cell surface of immature DC	110
13.3.	CD1a partitions to detergent-resistant membrane microdomains, which are necessary for efficient CD1a-restricted exogenous antigen presentation	111
13.4.	CD1a cartoon I - CD1a raft association, partners and dependency	112
13.5.	Tail-deletion does not change the subcellular distribution of CD1a	115
13.6.	CD1a TD shows normal surface expression, internalization and recycling	116
13.7.	CD1a TD efficiently presents sulfatide to T cells	117
13.8.	CD1a surface expression is not affected by inhibition of the clathrin-mediated pathway	119
13.9.	CD1a follows a Rab22a-dependent recycling pathway	120

13.10. CD1a is internalized and recycled by an ARF6-dependent pathway 121
13.11. CD1b does not accumulate in ARF6-Q67L-positive enlarged vesicles 122
13.12. CD1a cartoon II - CD1a mimics MHC class I trafficking 123
13.13. CD1a is redirected to lysosomes by providing residues of CD1b cytoplasmic tail . . 124
13.14. Marked intracellular colocalization between CD1a and sulfatide 126
13.15. WT CD1a transfectants present sulfatide more efficiently than CD1aab chimeric transfectants 128
13.16. CD1a:CD1b chimeras have shorter-lived ability to stimulate sulfatide-specific T cells than WT CD1a molecules 129
13.17. CD1a cartoon III - the view of CD1a trafficking and antigen-presentation to date . 130
14.1. Gamma-delta T cell-activation is increased by MRP5 overexpression 142
14.2. MRP5 overexpression increases resistance to ABC transporter blockers 143
14.3. Specific silencing of MRP5 visualized by CLSM 144
14.4. Specific silencing of MRP5 alters TCR gamma-delta cell-responses 145
14.5. Blocking of MRP5 by specific drugs decreases IPP transport over membranes . . . 146
14.6. IPP transport by MRP5 is ATP-dependent and can be competed by cold IPP . . . 147
14.7. MRP5 colocalizes partially with EE and ER 149
14.8. CLSM image analysis and quantification of MRP5 colocalization 150

List of Tables

5.1. Antibodies, human 31
5.2. Antibodies, mouse 34
5.3. Cell lines, human 34
5.4. Cell lines, mouse 35
5.5. Clones, human 35
5.6. Hybridomas, mouse 36
5.7. Compounds 36
5.8. Soluble recombinant birA-tagged human CD1 proteins 38

9.1. Clinical characteristics of the Lucerne patients 51
9.2. Clinical characteristics of the Bruderholz patients 53

11.1. Clonal brightness 64
11.2. DIMER blocking 68
11.3. The structure of a new analogue of alpha-galactosylceramide 71
11.4. The structures of new sphingolipid analogues 73
11.5. The structures of new aza- and thiasphingosines 76
11.6. The structures of single tail piperidinones 79
11.7. The structures of double tail piperidinones 80
11.8. Median lethal concentration of piperidinones 81

13.1. CD1a chimeras – tail to traffic twists 113

14.1. Gamma-delta T cell inhibition by transporter blockers 139
14.2. MRP4/5 expression in gamma-delta T cell-stimulatory cell lines 141

A Short History of Non-classical Immunology

The scientific field examining the body's defense or 'immune' (from medieval Latin *immunis* meaning untouchable or tax-exempt) system is called immunology. It originated from microbiology and medicine, especially from studies on the causes of immunity (resistance) to disease. During the plague of Athens in 430 before Christ, Thucydides committed to writing the earliest known mention of immunity in history. He noticed that persons having recovered from a previous insult of the disease could nurse the sick without contracting the illness a second time.

More than a millennium later, in 1718, Lady Mary Wortley Montagu, the wife of the British ambassador to Constantinople, observed the positive effects of variolation on the native Ottoman population and had the technique performed on her own children. This is the first report, in her "Turkish Embassy Letters", of an archetype of a today's vital immunological application namely vaccination.

The principle of vaccination was first demonstrated in 1798 on the example of smallpox by Edward Jenner. Since these first immunological trials on men the concept of immunology developed throughout the 19th and 20th centuries into a scientific theory. The work of Edward Jenner, Jakob Henle, and Ignaz Semmelweis inspired the scientist of the first golden age of immunology (1880-1910) to develop their theories. This era of Louis Pasteur, Robert Koch, Emil von Behring, and Paul Ehrlich was boosted by the hunt for vaccines and thus immunology was closely linked to microbiology.

Experiments with scorpion venom by Pierre-Louis Moreau de Maupertuis in the 18th century where he observed that certain dogs and mice were immune to this venom and other observations of acquired immunity were later exploited by Louis Pasteur in his development of vaccination and his proposed germ theory of disease. Pasteur's theory was in direct opposition to contemporary theories of disease, such as the miasmatic theory (from Greek *miasma* meaning a noxious atmosphere or influence). Only in 1981, when Robert Koch formulated his proofs, known as Koch's postulates and awarded a Nobel Prize in 1905, microorganisms were confirmed as a cause of infectious diseases. Viruses were recognized as human pathogens in 1901 owing to the discovery of the yellow fever virus by Walter Reed.

The second golden age of Immunology was preceded by and based on the discovery of antibodies as by Paul Ehrlich (jointly awarded a Nobel Prize in 1908 with the founder of cellular immunology, Elie Metchnikoff) who produced antibodies to the plant toxin ricin and formulated his side-chain theory already around 1900 and by Jules Bordet who found antibodies to red blood cells. With the proof by Karl Landsteiner that antibodies could be formed to recognize material that never existed in nature, immunology was freed from its microbiology ties. Thus antibodies were the advent of the second golden age of immunology (1955-1975) when nature's defense of life began to be unraveled. The second golden age initiated with the enunciation of the clonal selection theory by Frank Macfarlane Burnet and peaked by the invention of the technique and the production of the first monoclonal antibodies by Georges Köhler and César Milstein. Still it lasted till 1985 before the T cell receptor genes were identified. Slowly the dogma of the classical immune system with its components (e.g. the innate and adaptive immunity) was built up.

In a protein world lipids were reduced to structural components of living matter mainly without function especially in the sophisticated immune system. T cells specifically recognizing CD1 molecules were first described in 1989 [Porcelli et al. 1989] followed by further reports about alpha-beta and gamma-delta T cells restricted to different CD1 molecules; still the antigen remained elusive. Studies concluded that T cell responses to *Mycobacterium tuberculosis* were independent of major histocompatibility complex molecules however they depended on CD1b expression by antigen-presenting cells [Porcelli et al. 1992]. Nevertheless the nature of the antigen presented by CD1 molecules was identified not until 1994 when mycolic acid from *Mycobacterium tuberculosis* was proven the first lipid to be antigenic [Beckman et al. 1994]. Finally, when in 1997 the crystal structure of the first CD1 (mouse CD1d1, corresponding to human CD1d) molecule was solved

[Zeng et al. 1997], it revealed the mechanism by which lipids stimulate specific T cells.

In the last two decades non-classical components of the immune system were perceived and during the very last decade started to be appreciated functionally [De Libero & Mori 2009 a]. To these components belong the CD1 molecules and the non-peptidic antigen-specific T cells. In contrast to most classical components being highly polymorphic, the non-classical antigen-presenting machinery seems to be less or even non-polymorphic. That fact is to open a whole new world for vaccines and medical treatment because such a therapy could be applied to the whole world's population. Therefore clarification of how lipids become immunogenic and activate T cells is crucial to guide and to influence the body's defense mechanisms to its best by lipids being designed for a desired function.

Part I.

Introduction

1. The immune system

The immune system of the body is built on three defense lines against infections and alterations of self like cancer. The skin and other structural components of the body provide a first, physicochemical barrier to invasion by pathogens and are able to isolate malignant cells in enclaves as well as to exclaves. The second line of defense is provided by the innate immune system taking immediate action whereas the third defense line, the adaptive immune system, is lagging few to several days before the onset of a response.

Both the innate and the adaptive immune system contain so called humoral, acellular and cellular integrants that depend on glycosylation [Arnold et al. 2007; Rudd et al. 2001]. Cells of the innate system comprise macrophages and natural killer (NK) cells which rapidly kill their targets upon encounter after engulfment or from outside, respectively. The complement system, a cascade of proteins attacking invaders, is a major acellular innate defense mechanism and functions in cooperation with antibodies mostly and most efficiently of the IgM subtype. In contrast to the rather non-specific response of the innate, the adaptive immune system reacts highly specifically by its major lymphocyte cells, the T and B cells. B cells originate from the bone marrow and are the source of antibodies in the body. Conventional and unconventional T cells arise in the thymus, where they are positively and negatively selected for their T cell receptor (TCR) to establish restriction and avoid autoimmunity, respectively [Weinreich & Hogquist 2008]. Emigration from this developmental site is tightly controlled by molecules like the sphingolipid receptor 1 and the Krüppel-like transcription factor 2 [Carlson et al. 2006].

One major difference of the adaptive to the innate immune system is the establishment of memory to react to a threat with high specificity in a faster manner - in hours rather than days needed for a primary response. Memory is seen in the B and T cell-compartment [Mitchell 2008; Rensing et al. 2009]. Nowadays, generation of memory is employed in medicine by vaccination to protect humans before they get infected [Castellino et al. 2009].

Lately described unconventional T cells like the natural killer T (NKT) and $\gamma\delta$ T cells seem to phenotypically and functionally belong to both the innate and the adaptive immune system and to be able to bridge between the two systems. These T cells are known to immediately respond to pathogens and recently induction of glycolipid-specific memory T cells has been observed after vaccination in cattle [Nguyen et al. 2009]. In combining the advantages of the innate and the adaptive immune system by a fast and highly specific (memory) response, respectively, these cells may be of crucial importance for the body's immune defense. Still many aspects and mechanisms by which these cells function are poorly understood and merit more scientific attention.

2. Non-peptidic molecules

2.1. Lipid evolution and repertoires

The abundance and diversity of lipids in microbes are not ignored by the immune system. Lipid-reactive T cells complement and extend classical major histocompatibility complex (MHC)-restricted peptide-specific T cells. The non-classical T cells provide the second, lipid-based arm of the immune system that pathogens have to evade to be successful.

The lipidome of a single mammalian cell comprises more than a thousand different lipids [van Meer 2005] and their mass equals the one of proteins in membranes. Most of them are built by a polar head and hydrophobic tail(s). Their life-sustaining and thus most vital function for a cell lies in a 3 nm hydrophobic boundary delimiting the environment. These membranes dynamically organize both across the lipid bilayer (lipid asymmetry) and in the lateral dimension (lipid domains) [Somerharju et al. 2009]. Due to their intrinsic physical phase behavior and their interactions with membrane proteins, they create unique compositions and multiple functionalities of individual membranes of either compartments or even within a compartmental membrane [van Meer et al. 2008].

By means of regulating anabolic and catabolic enzymes, of positioning translocases, and of activating lipid transporters, cells can manipulate their lipidome. Cellular lipidomics [van Meer 2005] allocates a framework for understanding the role of lipids in time and space of a cell's life.

Small genetic differences exist between higher mammalian species. Yet they are crucial and the major known ones include the accumulation of subcutaneous fat, the growth of the brain and the connectivity of neurons. All these achievements involve lipid metabolism, nevertheless lipids are rarely mentioned in discussions about (human) evolution, because fat leaves no fossils [Horrobin 1999]. Biochemistry of lipids and evolution are closely connected and will benefit from recent advances in lipidomics even without historical traces.

The lipidomes of microbes can be similarly complex or relatively simple but have to fulfill the same three general functions: energy storage, membrane matrix supply and signaling as first and second messengers. Given the variety of lipids and their tightly regulated spatial and temporal occurrence, it is not surprising that one can find lipids being specific for a cell or microbe or even their (ephemeral) state (e.g. caused by viral infection or tumorigenesis). This fact is exploited by the immune system to discriminate self from foreign or housekeeping from danger signals emerging from the lipidome. Additionally, the discovery of new disease-related genes has proven the importance of lipid-related proteins, such as lysosomal hydrolases, ATP-binding cassette (ABC) transporters or lipid binding proteins in both lipid physiology [Kolter & Sandhoff 2009; van Meer et al. 2008] and immune responses [De Libero & Mori 2006, 2009 a].

2.2. Lipid localization and metabolism

Glycerophospholipids (phosphatidylcholine, phosphatidylethanolamine, phosphatidylserine, phosphatidylinositol and phosphatidic acid) are the major structural lipids in eukaryotic membranes. Their hydrophobic backbone is built of a diacylglycerol with saturated or cis-unsaturated fatty acyl chains of varying lengths. Another class of structural lipids are sphingolipids containing a ceramide backbone [van Meer & Hoetzl 2009]. The major mammalian sphingolipids are sphingomyelin and glycosphingolipids (mono-, di- or oligosaccharides) based on glucosylceramide and occasionally galactosylceramide with saturated or trans-unsaturated acyl chains. Sterols provide the majority of non-polar lipids of cell membranes with cholesterol predominating in mammals.

Hydrolysis of glycerolipids and sphingolipids is used to produce parallel series of messenger lipids as lysophosphatidylcholine, lysophosphatidic acid, sphingosylphosphorylcholine, sphingosine, and sphingosine-1-phosphate with only one aliphatic chain, that can readily leave membranes and thereafter signal through related membrane receptors, and phosphatidic acid, diacylglycerol, ceramide-1-phosphate, and ceramide, that remain in the membrane and can recruit cytosolic proteins. Synthesis of structural lipids is spatially restricted inside eukaryotic cells. Compartmentalized lipid metabolism is the first determinant of the unique local compositions of organelles. When during evolution eukaryotes started to synthesize sphingolipids and sterols, physical differences between these and the glycerophospholipids enabled cells to segregate lipids within membranes. Secluding this event to a dedicated organelle, the Golgi, allowed creation of membranes of different lipid composition, notably a thin, flexible endoplasmic reticulum (ER), built of glycerolipids, and a robust plasma membrane containing at least 50% sphingolipids and sterols. Besides sorting membrane proteins, sphingolipids obtained key positions in cell physiology and signaling [van Meer & Hoetzl 2009].

The ER synthesizes the bulk of cellular lipids, structural (phospholipids and cholesterol) and non-structural ones (triacylglycerol esters and the sphingolipid precursor ceramide). Specific lipid biosynthetic enzymes are substantially enriched in mitochondria-associated membranes, a ER subfraction, pointing to further subcompartmentalization of lipid synthesis in the ER. The ER, situated at the beginning of the secretory pathway, displays low concentrations of sterols and complex sphingolipids resulting in loose membrane packing. This permits efficient membrane insertion and transport of newly synthesized lipids and proteins. The Golgi is specialized in sphingolipid and glycosphingolipid synthesis, primarily destined for export to the plasma membrane. The Golgi and its network constitute the logistic center of a cell, sorting lipids and proteins both anterogradely and retrogradely. The production of sphingolipids may be linked to the vesicular transport [Murphy et al. 2009; Zehmer et al. 2009] between different organelles through lipid rafts. Secretory processes may be controlled by the local diacylglycerol levels that in turn are regulated by the Golgi-specific cholinephosphotransferase. Plasma membranes, built to resist mechanical stress, are enriched in sphingolipids and sterols that grant packing at higher density than glycerolipids. Even though the plasma membrane does not synthesize its own structural components, it significantly participates in synthesis of total cellular sphingomyelin from ceramide by a sphingomyelin synthase. The plasma membrane has important functions as mechanical barrier and as signaling interface between intra- and extracellular space. Signaling involves many lipid-synthesizing and -degrading reactions to drive the messenger cascades described for the plasma membrane. Endosomal membranes of the endocytic pathway are similar to the plasma membrane during their early stage, but upon maturation to late endosomes (LE) and internal acidification, there is a decrease in sterols and phosphatidylserine versus a drastic increase in bismonoacylglycerophosphate. The latter works in multivesicular body formation, fusion processes and sphingolipid hydrolysis. Differentially phosphorylated phosphoinositides identify endocytic membranes and allow them to recruit cytosolic proteins involved in vesicle trafficking [Haucke & Di Paolo 2007]. The overall amounts of signaling lipid mediators are negligible compared to total membrane lipids, nevertheless they are crucial for cell functions. Topological and temporal regulation of these signaling systems are largely unknown.

2.2. Lipid localization and metabolism

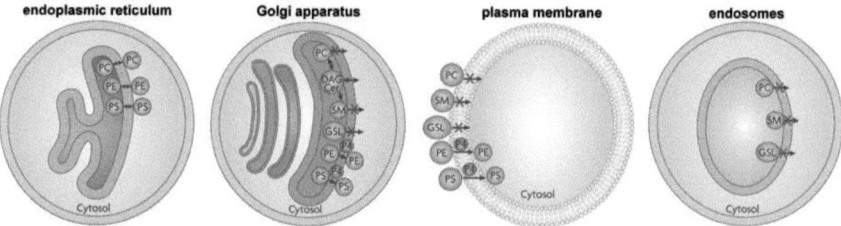

Figure 2.1
Spatial lipid organization to form membrane asymmetry and domains. During the travel of lipids from the endoplasmic reticulum through the Golgi, plasma membrane and endosomes, intrinsic physicochemical properties and lipid transporters determine the phospholipid distribution across the bilayer leaflets. The ER membrane exhibits a widely symmetric lipid bilayer due to intrinsic equilibration of phospholipids. In the Golgi, P4 ATPases (P4) translocate phosphatidylserine (PS) and phosphatidylethanolamine (PE) to the cytosolic side of the bilayer leaflet. Sphingomyelin (SM) is formed by sphingomyelin synthase from ceramide (Cer) on the luminal side. Neither phosphatidylcholine (PC) nor sphingomyelin are transported to the cytosolic face but stay resident in the luminal leaflet. Asymmetry in the Golgi is achieved by specific transport of phosphatidylserine and phosphatidylethanolamine as well as lack of transport of sphingomyelin and phosphatidylcholine. During sphingomyelin synthesis, phosphatidylcholine is converted to diacylglycerol (DAG), which freely diffuses across bilayers. Diacylglycerol can serve as a substrate for the Golgi cholinephosphotransferase producing phosphatidylcholine. At the plasma membrane, P4 ATPases transport phosphatidylserine and phosphatidylethanolamine to the cytosolic face, with little or no transport of phosphatidylcholine or sphingomyelin under basal conditions. This homeostatic equilibrium is sensitive to disruption by scramblase and/or by inhibition of P4 ATPases. Endosomes were shown to restrict phosphatidylcholine, sphingomyelin, and glycosphingolipids (GSL) to the luminal leaflet without any specific transport mechanisms. Adapted from [van Meer et al. 2008].

Further organelles include multivesicular bodies, originating from inward budding of endosomes, and mitochondria, that have evolved from bacterial origin as they contain microbial lipids [Griffiths 2007]. They synthesize large amounts of lipids like lysophosphatidic acid, phosphatidic acid, cardiolipin from phosphatidylglycerol, and phosphatidylethanolamine to be exported by decarboxylation of phosphatidylserine. Negatively charged lipids such as fatty acids, phosphatidylglycerol, and phosphatidic acid can rapidly flip when their charge is neutralized by low pH, but they can effectively be accumulated on one side of the membrane bilayer by a pH gradient. In contrast to the ER bilayer, the Golgi, plasma, and endosomal membranes are asymmetrically distributed with sphingomyelin and glycosphingolipids in the non-cytosolic (luminal) leaflet and with phosphatidylserine and phosphatidylethanolamine in the cytosolic leaflet. This asymmetrical distribution of lipids in membranes (see Figure 2.1, p.11) has important functional consequences, in signaling, vesicule formation and traffic, and in keeping the identity of cellular compartments. How this asymmetrical distribution impacts on lipid immunogenicity is not know to date.

Translocation of lipids can occur by non-specific lipid transporters within membranes of high phospholipid synthesis capacity. This transport appears to depend on membrane protein subset and is adenosine triphosphate (ATP)-independent. This equilibration property is lost in non-ER membranes of eukaryotes and in the outer membrane of prokaryotes, due to changes in lipid and/or protein composition. On the other hand, glycolipid precursors for protein glycosylation (oligosaccharides with a prenyldiphosphate backbone) are translocated from the cytosol across the ER and the bacterial inner membranes by dedicated transporters.

Deletion of certain P4 ATPases not only affects lipid flipping from the non- to the cytosolic hemileaflet but also blocks vesicular traffic, suggesting a role for lipid asymmetry in membrane bending and vesicle budding. Other membrane-bound lipid transporters include ABC transporters, that cardinally work in the opposite direction of P4 ATPases and moreover export lipids by expelling them from the membrane rather than acting as flippases, and the plasma membrane scramblase, which randomizes plasma membrane phospholipid distribution independently of ATP. The nature of the scramblase is still poorly understood. For many ABC transporters, specific pathologies have been genetically linked, but the subjacent mechanisms remain disputed as do the substrates for the transporters.

2.3. Lipid cellular trafficking

Every organelle specifically contains some lipids that were produced elsewhere and acquired by non-random transport mechanisms. The plasma membrane, endosomes and lysosomes depend completely on lipid transport occurring by various mechanisms. The major membrane transport pathway between cellular organelles is through the budding and fusion of membrane vesicles in the secretory and endocytic pathways. Alternatively, lipids enter and leave organelles not connected to the vesicular transport system, such as mitochondria and peroxisomes, with the help of both soluble and membrane-bound proteins. After exit from the ER several lipids become restricted to the exoplasmic leaflet of the bilayer, lose access to the cytosolic leaflet, and require vesicle-mediated transport. Even the access to retrograde vesicles seems to be locked for certain lipids preferentially trafficking in anterograde vesicles.

The lipid raft hypothesis proposes that preferential interactions between lipids generate domains of specific lipid compositions to drive the sorting of membrane proteins [Sengupta et al. 2007; Simons & Ikonen 1997; Yuyama et al. 2003]. Sphingolipids concentrate in rafts that already exist in the ER membrane and persist throughout intracellular pathways, as single endosomes show non-randomly distributed sphingolipids, indicative of domains with high sphingolipid content. Lactosylceramide is preferentially endocytosed by caveolae and transported from the endosomes to the Golgi. This indicates that the endocytic recycling system segregates lipids by similar mechanisms to those of the secretory pathway. There are hints that lipid–lipid interactions regulate the initial step of caveolar endocytosis. In general, both protein and lipid assemblies on donor and acceptor membranes are important for membrane–membrane recognition and interactions that facilitate the non-vesicular lipid transfer processes. This combinatorial recognition process has the intrinsic ability to provide spatial and temporal cues that constitute a 'chemical roadmap' for vectorial lipid transport. Each membrane, each hemileaflet and each segregated domain contains only a subset of the lipidome. The identification of these domains and the elucidation of the relationships between physical state and function define the current limits of our understanding. For example, phase separations may be driven by the loss of certain lipid components together with the addition of new ones, as occurs in consequence of the activity of lipid kinases, phosphatases and hydrolases during signaling reactions, or by changes in Ca^{2+} concentration, or by the aggregation of immune receptors [Sengupta et al. 2007]. Proteins can show a clear preference for a particular lipid phase or even induce it. Most membrane-bound peptides partition out of the liquid-ordered (raft) into liquid-disordered domains; exceptions are peptides of the caveolin-1 scaffolding domain and at least some glycosylphosphatidylinositol-anchored proteins.

2.4. Phosphorylated metabolites

The main source of phosphorylated metabolites in eukaryotes, plants, and archaea is the cytosolic mevalonate pathway and its downstream products [Morita et al. 2007]. The alternative 2-C-methylerythritol 4-phosphate (MEP) pathway is used in most bacteria (with the exception of gram-positive cocci), apicomplexan protozoa, and chloroplasts. HMG-CoA reductase, which is the target of statin drugs and of negative feedback regulation by downstream products, is the key regulatory enzyme of the mevalonate pathway and its MEP pathway ortholog Dxr [Gober et al. 2003]. The enzyme farnesyl pyrophosphate (FPP) synthase represents the initial step to the mevalonate pathway downstream compounds and converts isopentenyl pyrophosphate (IPP) (and dimethylallyl pyrophosphate (DMAPP)) to geranyl and farnesyl pyrophosphate. FPP synthase is inhibited by aminobisphosphonates and alkylamines [Das et al. 2001; Dunford et al. 2001; Kunzmann et al. 2000; Thompson et al. 2006]. Loss of FPP and geranylgeranyl pyrophosphate prevents membrane anchoring of many signaling proteins, causing signaling defects and in some cases cell apoptosis.

2.5. Foreign and self non-peptidic antigens

Bacterial phosphorylated metabolites were the first antigens identified for TCR $V\gamma 9/V\delta 2$ cells and they were isolated from *Mycobacterium tuberculosis* lysates [De Libero 1997]. Later more non-peptidic, low molecular weight compounds mostly containing an acid-labile pyrophosphate group with various backbone structures were found stimulatory [Buerk et al. 1995; Constant et al. 1994; Eberl et al. 2003; Tanaka et al. 1995].

Mycobacterium tuberculosis extracts were also instrumental in isolating the first CD1(b)-restricted T cells [Porcelli et al. 1992]. The stimulatory antigens were later discovered to be mycolic acids [Beckman et al. 1994]. CD1a-restricted antigens, such as didehydroxymycobactin [Moody et al. 2004], and CD1c-restricted antigens, such as mannosyl phosphomycoketide [Moody et al. 2000], were found in *Mycobacterium tuberculosis* sonicates, too. To date, mycobacte-

2. Non-peptidic molecules

riaceae and other microbes of the actinomycetales order are the most productive source of group 1 CD1-presented antigens [Willcox et al. 2007]. Further CD1b-restricted antigens as lipoarabinomannan [Sieling et al. 1995], a phosphatidylinositol mannoside, and 2-palmitoyl or 2-stearoyl-3-hydroxyphthioceranoyl-2'-sulfate-α-α'-D-trehalose (Ac$_2$SGL) [Gilleron et al. 2004], a mycobacterial diacylated sulfoglycolipid, have been found and are more closely related to group 2 CD1-presented antigens. Indeed, a synthetic phosphatidylinositol dimannoside has been co-crystallized with mouse CD1d [Zajonc et al. 2006]. The most prominent structural feature of CD1d-bound antigens, namely glycosphingolipids, seems to be their α-anomerically attached sugar headgroup contrasting mammalian glycosphingolipids containing β linkages. In addition to alpha-galactosylceramide, the archetype invariant natural killer T (iNKT) cell ligand, other antigens as α-linked sphingolipids from *Sphingomonas* [Kinjo et al. 2005] and α-linked galactosyl diacylglycerol from *Borrelia burgdorferi* [Kinjo et al. 2006] were shown to activate iNKT cells.

Most non-peptide-specific T cells present a memory-like response after antigen encounter. This has been ascribed to continuous, homeostatic TCR triggering by self antigens as IPP for $\gamma\delta$ T cells or the disputed isoglobotrihexosylceramide (iGb3) for iNKT cells. Additionally, self ligands may be necessary for proper folding as well as stable assembly and trafficking of CD1 molecules. Finally, autoimmune diseases like atherosclerosis (ATH) and multiple sclerosis may be connected to T cells recognizing self non-peptidic compounds [De Libero & Mori 2007 a].

3. Antigen-presenting molecules (APM)

On cells loaded with non-peptidic compounds antigen-presenting molecules (APM) allow crosstalk with a variety of T cells.

3.1. CD1

Immunization of mice with human thymocytes led to the discovery of a highly specific antibody recognizing 85% of the human thymic cell population [McMichael et al. 1979]. This antibody (NA1/34) was later shown to bind CD1a of the CD1 proteins. These molecules are able to bind and present lipid antigens to T cells.

3.1.1. CD1 evolution and morphology

The CD1 genes are highly conserved in mammalian species. Their existence has been known for more than two decades [McMichael et al. 1979] but only during the last decade the functions of CD1 proteins are being appreciated [Vincent et al. 2003]. CD1 proteins, even though being structurally related to the MHC class I family, have evolved to bind and to present lipidic antigens to T cells instead of peptides.

The human CD1 locus spans $\approx$ 175 kilobases on chromosome 1 and is paralogous to chromosome 6 [Calabi & Milstein 1986]. The mouse CD1 locus, located on chromosome 3, is also paralogous to the MHC locus on chromosome 17. In man, the CD1 genes encode five distinct isoforms [Calabi & Milstein 2000] subdivided into group 1 (CD1a, CD1b and CD1c), group 2 (CD1d) and group 3 (CD1e) based on amino acid sequence similarity. Molecular phylogenetic analysis of CD1 proteins from a variety of species representing several major mammalian orders using comparison of gene sequences by alignment points to a common ancestor probably as early as before the end of the Mesozoic era [Dascher & Brenner 2003]. In non-mammalian vertebrate species existence of CD1 genes has not been evidenced so far, although not being extensively searched for.

In addition to differences in sequence, there is a growing body of evidence that the group 1, group 2 and group 3 CD1 isoforms also have distinct functions in the host immune system. The group 1 CD1 (CD1a, CD1b, CD1c) proteins are more involved in presentation of foreign lipid antigens, whereas group 2 (CD1d) recognize self-lipid antigens and play a more immunoregulatory role, although there could be some degree of overlap in these two functions. Hence, the absence of group 1 CD1 genes in the mouse could have important consequences for evaluating the contribution of CD1 antigen presentation in various human disease models.

The deletion of the group 1 CD1 genes from the mouse genome was first proposed by Bradbury and Milstein [Bradbury et al. 1990] and confirmed after completion of the mouse and human genome sequencing [Dascher & Brenner 2003]. Compared with humans, who have single copies of the five known CD1 isoforms, other species have variations in the number of CD1 genes, as well as variations in the specific isoforms that are present in their genomes. The heterogeneity of the CD1 gene family between species raises certain fundamental questions regarding the function of

3. Antigen-presenting molecules (APM)

the individual isoforms. A conceptual breakthrough in understanding the potential role of the various human CD1 isoforms (CD1a, -b, -c and -d) has emerged with the appreciation that each isoform has a unique trafficking pattern through the endosomal system of antigen-presenting cells (APC). Lack of certain CD1 genes in other species than human might be compensated for by altered intracellular trafficking of the available CD1 isoforms leading to 'functional' homologs of missing CD1 isoforms ('traffic' hypothesis [Dascher & Brenner 2003]). In line with this assumption, mouse CD1d markedly colocalizes with lysosome-associated membrane glycoprotein 1 (LAMP-1), a marker for lysosomes and LE, whereas human CD1d does so only partially. According to the 'traffic' hypothesis, selective pressure gave rise to mutations in the cytoplasmic tail changing the binding of adaptor proteins (AP) and thereby the trafficking pattern. Similar convergent functional evolution can be seen for the guinea pig CD1b3 isoform that resembles human CD1b in its sequence but human CD1a in its intracellular trafficking due to a single nucleotide transition from the tyrosine in the tyrosine-based endosomal sorting motif to a cysteine [Hiromatsu et al. 2002].

Selection pressure may not only change the cytoplasmic tail to broaden the compartment survey system of CD1 proteins but additionally act on the antigen binding groove to allow a larger variety of antigens to be loaded and presented ('groove' hypothesis). This will increase a species' fitness by the ability to present so far inaccessible antigens when encountering pathogens. This has been confirmed by crystal structures of the CD1 isoforms revealing characteristic modifications of the antigen binding groove, as f.i. the extra tunnel ('T') or portal at the distal end of the C' pocket in human CD1b [Zajonc & Wilson 2007].

Evolutionary modifications to the CD1 gene family as assumed by the 'traffic' and 'groove' hypotheses could be simultaneous or temporally distinct incidents. A sequential scenario has been favored, in which mutations of the cytoplasmic tail change the intracellular trafficking followed by adaptation of the binding groove to the new environment (see Figure 3.1, p.17), because selective pressure on the groove is depending on and exerted by differential lipid compositions of the intracellular compartments. This is in accordance with the fact that optimal loading of lipid antigens with longer acyl chains, which are actively sorted inside a cell based on the chain length [Mukherjee et al. 1999], requires meeting of the CD1 molecules in LE [Moody et al. 2002, 2003].

Natural selection pressure shaping the immune system of the fittest to survive can be seen in the evolution of APM genes by their sequences and by their crystallographic structures even more conclusively.

The first solved crystal structure of CD1, namely mouse CD1d1 [Zeng et al. 1997], resembles the one of MHC class I molecules and also associates with β_2 microglobulin (β_2m) by means of conserved motifs in the α3 immunoglobulin like domain [Tysoe-Calnon et al. 1991]. Both CD1 and MHC class I molecules are anchored to the cell membranes by a transmembrane domain and a cytoplasmic tail attached to the α3 heavy chain domain. In the membrane distal part of the molecules, above the α3 domain and the β_2m, two α helices (α1 and α2) lie atop a floor of six anti-parallel β-pleated sheets. The main difference is seen in the antigen-binding site of MHC class I and CD1 molecules with the shallow, rather hydrophilic groove of MHC class I compared to the deep, hydrophobic groove of CD1. The several hydrophobic channels of the latter groove allow binding of hydrophobic structures as the acyl chains of lipids.

CD1 molecules are found to have either two or four hydrophobic pockets termed A', C', F', and T', with A', C', and F' corresponding roughly to the eponymous MHC class I structure

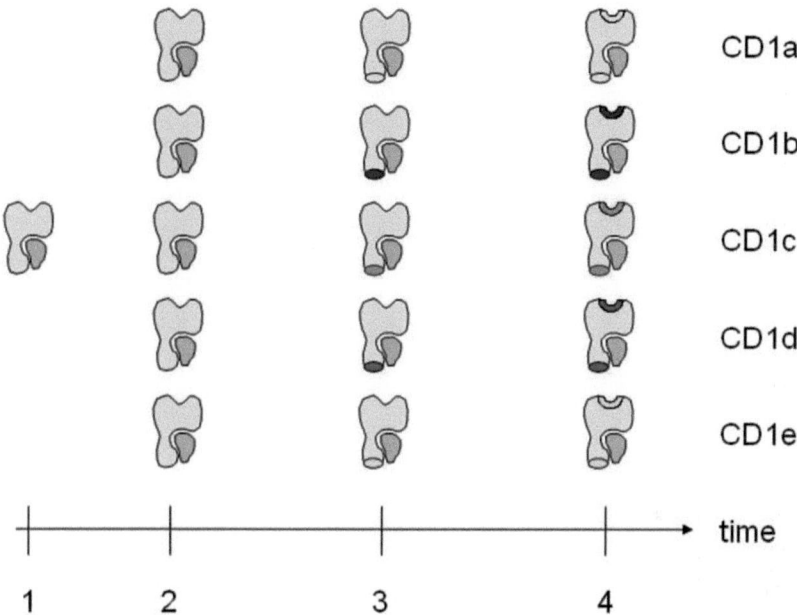

Figure 3.1
Evolutionary hypothesis of the human CD1 multigene family. A primordial CD1 (1) arises by duplication and diversification from an MHC class I gene into a lipid-binding molecule with the capacity to present the bound lipid to T cells and to traffic to some or all endosomal compartments. Gene duplications (2) occur creating multiple CD1 isoforms. Mutations in the cytoplasmic tail (3) of different isoforms result in differential endosomal trafficking of the respective isoform ('traffic' hypothesis). Exposure to selected subparts of the endocytic system and the prevailing physiological conditions in it as well as to quantitatively and qualitatively distinct subsets of lipids provide selective pressure to mutate the groove (4) for optimal binding of lipid antigens ('groove' hypothesis). This optimized trafficking and lipid antigen binding confers evolutionary advantage to the host. Adapted from [Dascher & Brenner 2003].

counterparts [Moody et al. 2005]. T' is a tunnel uniquely found in CD1b where it connects the A' and F' pockets; additionally, CD1b has an extra opening portal at the end of the C' channel [Gadola et al. 2002 a]. Whereas the A' and F' pockets of CD1b are required for optimal antigen presentation [Niazi et al. 2001], the role of the C' pocket is not fully understood. Because the A' and C' pockets seem to be partially closed, the main lipid entry site is the F' portal providing access to the A' and C' channels as well. Ionizable residues in the F' portal (of CD1b, CD1c, and CD1d isoforms) might lead to pH-controlled partial unfolding of the α helices [Koch et al. 2005]. CD1a is lacking these residues as it does not travel to the late endosomal and lysosomal compartments confirming a potential low pH-induced 'lipid receptive' state of the other CD1 isoforms. Furthermore, the F' portal plays a crucial role in positioning the hydrophilic lipid headgroups by a hydrogen bonding network. The sphingosine base of the antigen backbone commonly binds to the F' pockets leaving the acyl chain to adapt around a structural pole in the A' pocket. Even though the nature and existence of the four pockets varies among the isoforms, the acyl chain length allowed is relatively similar with the exception of CD1b wherein longer lipid chains can fit due to the T' connected A' and F' channels. CD1a shows the shallowest and smallest antigen-binding groove, whereas CD1b with its additional C' and T' channels attains the largest cavity. CD1d has an intermediately sized total volume but CD1d and CD1b can bind shorter lipid structures with the help of a 'spacer' lipid [Garcia-Alles et al. 2006; Wu et al. 2006; Zajonc et al. 2005 b] filling the else empty space. To date, no crystal structure of CD1c has been published and thus binding of antigens by CD1c is solely predictive. The more perpendicular insertion of CD1-presented antigenic lipid, in contrast to the longitudinal orientation of peptides on MHC molecules, exposes the lipid headgroup more towards the solvent and TCR.

3.1.2. CD1 tissue expression and cell distribution

Presentation of antigens by APM is not only temporally but also spatially controlled by tissue- and cell-specific expression of e.g. CD1 proteins. In accordance with their function to present lipids and glycolipids to T cells, the CD1a, CD1b, and CD1c proteins are expressed primarily on APC [Dougan et al. 2007], including DC and B cells [Porcelli & Modlin 1999]; CD1d has a broader distribution on lymphoid and myeloid lineage cells [Exley et al. 2000]. In the thymus, group 1 CD1 proteins (CD1a, CD1b and CD1c) are mainly expressed on CD4$^+$CD8$^+$ double-positive (DP) cortical thymocytes and CD1d expression on thymocytes gradually decreases during development with downregulation from cortical to medullary thymocytes and absence on naïve peripheral T cells, albeit T cells retain the capacity to re-express CD1d upon activation [Exley et al. 2000]. Furthermore, CD1a is used as an identification marker of Langerhans cells [Pena-Cruz et al. 2001; Pena-Cruz et al. 2003].

CD1d is widely distributed also outside the hematopoietic system and its expression on parenchymal (like hepatocytes, pancreatic, renal, and tonsillar cells) and endothelial cells (EC) contrasts group 1 CD1 molecules [Dougan et al. 2007]. Especially high expression levels of CD1d were found on vascular smooth muscle cells in all tissues [Canchis et al. 1993].

A major difference to MHC class II molecules exists as CD1 expression does change only marginally upon dendritic cell (DC) maturation [Cao et al. 2002; Sugita et al. 1996; van der Wel et al. 2003]. Therefore CD1-restricted antigens are efficiently presented by immature and mature DC contrasting the enhanced peptide-MHC class II presentation of mature DC. Nevertheless, upregulation of f.i. CD1d has been observed on hepatocytes after viral infection [Durante-Mangoni et al. 2004; Tsuneyama et al. 1998].

3.1.3. CD1 assembly, trafficking and loading

Akin their structure the assembly of CD1 proteins proceeds similarly to MHC class I molecules. Directed to the ER lumen by mRNA-contained leader sequences, CD1 precursors are recognized by the chaperones calnexin and calreticulin and thereafter ERp57 [Kang & Cresswell 2002 a]. β_2m binding is the *sine qua non* for most MHC class I proteins to exit the ER but a fraction of CD1d heavy chain is able to leave the ER without β_2m. The CD1 isoforms appear to have different requirements for ER exit as f.i. CD1b relies more on pairing with β_2m. After formation of the disulfide bonds in the heavy chain and finished glycosylation of the assembled complex, MHC class I molecules become loaded with peptides translocated by the transporter associated with antigen processing (TAP) transporter and facilitated by tapasin protein. Albeit there is evidence that CD1 complexes are loaded with endogenous lipid ligands, proteins working in analogy to TAP and tapasin have not been identified. The role in lipid CD1 loading of the ER-located microsomal triglyceride transfer protein (MTTP), that has been reported necessary for iNKT cell activation by CD1d [Brozovic et al. 2004; Dougan et al. 2005] and important for all group 1 CD1 isoforms in antigen presentation [Kaser et al. 2008], remains controversial as the mechanisms by which it affects exogenous antigen presentation and lysosomal recycling are elusive.

After exit from the ER, CD1 proteins follow a direct secretory route [Briken et al. 2002]. Nevertheless subfractions of CD1 molecules can be deviated from this route by binding to other molecules as proven for CD1d and invariant chain (Ii) and MHC class II [Jayawardena-Wolf et al. 2001; Kang & Cresswell 2002 b]. Passing by the Golgi the CD1 proteins reach the plasma membrane, maybe stabilized by endogenous lipids. Afterwards, they are differentially internalized into the endocytic system [Moody & Porcelli 2003]. Why they are detouring via the plasma membrane is still poorly understood. The prerequisite for CD1b, CD1c, and CD1d internalization and trafficking is their tyrosine-based motif in the cytoplasmic tail, that allows binding of adaptor proteins (e.g. AP-2 for CD1b internalization) and sorting by the clathrin-dependent pathway [Bonifacino & Traub 2003; Brigl & Brenner 2004]. Re-internalization seems to be a requirement for efficient antigen presentation as tail-deleted mutants of CD1 proteins show functional defects in T cell activation [Chiu et al. 2002; Jackman et al. 1998]. Without the tyrosine-based motif, CD1a still enters the endocytic system but takes a clathrin-independent route [Barral et al. 2008].

Humans CD1a recycles from the plasma membrane mainly in early endosomes (EE) [Sugita et al. 1999]. CD1b is routed through LE and lysosomes, residing in specialized lysosomes with MHC class II (MIIC compartment) [Sugita et al. 1996]. Adaptor protein 3 (AP-3) helps targeting proteins from sorting endosomes to lysosomes and has been found to bind to the cytoplasmic tail of CD1b [Sugita et al. 2002] exerting its targeting function with this protein, too. Human CD1b and mouse CD1d, being structurally the closest relatives, are located to LE and the lysosomal compartments under steady state conditions. AP-3 is essential as its deficiency leads to plasma membrane and early endosomal accumulation of CD1b as well as functional deficits including developmental defects in the CD1d-restricted mouse iNKT cell population. CD1c surveys both early and LE [Briken et al. 2000; Sugita et al. 2000 a] but due to the lack of AP-3 binding is mostly barred from lysosomal compartments. For the same reasons, CD1d only partially localizes to lysosomes.

CD1a and CD1c found in the endocytic recycling compartment are known to re-access the plasma membrane, and thus the cell surface for antigen presentation, in a adenosine diphosphate ribosylation factor (ARF) 6-dependent manner through the slow recycling compartment similar to MHC class I. Recycling mechanisms of CD1b and CD1d from LE and the lysosomal compartments back to the cell surface are less well understood. As some lipids need lysosomal processing to become immunogenic, immune evasion mechanisms invented by successful pathogens include blocking of the lysosomal recycling pathway, f.i. reported for herpes simplex virus I [Yuan et al. 2006]. The broad intracellular survey system constructed by the diverse compartmental trafficking of CD1 proteins tries to impede hiding of pathogens and their lipids [Moody & Porcelli 2003] from the immune system [Sugita et al. 2000 b]. Alterations during infections in the self lipid repertoire and its distribution [Mukherjee et al. 1999] are detected by the same means, namely variations in CD1 isoform trafficking [Sugita et al. 2007].

As lipids are embedded in membranes, before they can be accessed by dedicated enzymes, they must be extracted and made soluble. This is achieved by liftases and lipid-binding proteins which reside within the endosomal compartments. Several common ER lipids have been found associated with different CD1 isoforms [Cox et al. 2009; Garcia-Alles et al. 2006; Giabbai et al. 2005; Yuan et al. 2009]. ER-loading of endogenous lipids will overcome the energetically unfavorable empty hydrophobic cavities of the CD1 proteins and thus help in proper folding [Silk et al. 2008 b]. In some instances, endogenous spacer lipids are bound in order to fill the hydrophobic cavities and thus reduce the space within CD1 pockets available for the simultaneous binding of other lipids [Garcia-Alles et al. 2006; Wu et al. 2006; Zajonc et al. 2005 b]. Therefore, endogenous lipids and so called spacer lipids are central to the production of functional CD1 proteins.

Mainly lysosomal enzymes like α-D-mannosidase, α-galactosidase, and α-hexosaminidase process glycolipid antigens. The acidic environment participates in loading, by loosening the α helices around the antigen binding groove of CD1 molecules, and processing by hydrolyzing acid-sensitive chemical groups of the intact antigen, too. This has been strengthened by impact of diverse lysosomal storage diseases on iNKT cell development and stimulation [Gadola et al. 2006 a; Schuemann et al. 2007]. CD1e, the fifth member of the human CD1 proteins and the only one in group 3, stays intracellular and is slightly more polymorphic than its homologs. CD1e is implicated in processing of large glycolipid compounds thus allowing their binding to CD1b and formation of immunogenic complexes [de la Salle et al. 2005]. Whether CD1e has other important functions in lipid presentation remains to be investigated.

Lipid antigen processing has been reported for the headgroups but not for the tails so far. Not all antigens need processing [Manolova et al. 2006; Shamshiev et al. 2000] as f.i. alpha-galactosylceramide (αGC), the pan-iNKT cell activating glycosphingolipid, can be loaded onto CD1d exogenously and even in cell-free *in vitro* plate-bound assays.

The removal of the lipid antigens out of the membrane [Schulze et al. 2009] and their transfer into the antigen binding groove of CD1 proteins is assisted by a manifold of lipid transfer proteins [De Libero & Mori 2005 b; Kolter & Sandhoff 2009] including CD1e, saposins and GM2-activator protein [Kolter & Sandhoff 2005; Sandhoff & Kolter 2003]. The prosaposin is cleaved into four functional saposins (A-D) with individual roles in lipid metabolism and lipid antigen loading. They are involved in loading onto CD1b (saposin C) and mouse as well as human CD1d (saposin B) [Kang & Cresswell 2004; Yuan et al. 2007; Zhou et al. 2004 b].

Due to the nature of non-peptidic antigens the presentation machinery is not only depending on delivery by dedicated system from outside but also on intracellular proteins rendering the antigens accessible for loading and/or processing by freeing them from membranes and vesicles of self and foreign origin.

3.2. The antigen-presenting molecule for isopentenyl pyrophosphate and $\gamma\delta$ T cells

Despite intense investigations the putative APM for TCR $\gamma\delta$ cells remains mysterious. Attempts to visualize the APM by fluorescently labeled or crosslinkable antigen failed as the modified antigens lost activity. Blocking with antibodies to known APM like MHC class I/II or CD1 proteins were not successful [Lang et al. 1995; Morita et al. 1995 a]. The hypothesis, that the TCR $\gamma\delta$ recognizes the antigen without the need of an APM, arose from the independence of antigen processing by the APC but is unlikely as no TCR Vγ9/Vδ2 co-crystal with highly potent antigen could be obtained and moreover TCR Vγ9/Vδ2 cells need cell-cell contact to be stimulated by pyrophosphorylated antigens [Lang et al. 1995; Morita et al. 1995 a]. The requirement of cell-cell contact does not prove the existence of an APM, but it could suggest the dependency on costimulatory signals. Contradictory to such a dependency, the proliferative response of $\gamma\delta$ T cells to triggering with anti-CD3 alone was found equivalent to proliferation induced by anti-CD3 and anti-CD28 [Hayes & Love 2002].

Furthermore, antibodies to the $\gamma\delta$ TCR block T cell activation [Munk et al. 1990; Tanaka et al. 1994] and transfection of Vγ9/Vδ2 TCR can confer responsiveness onto a hitherto unresponsive cell [Bukowski et al. 1995]. Therefore, it is more likely that a specific, antigen-loaded APM is recognized by the TCR. The wide range of cell types that efficiently present phosphorylated metabolites to $\gamma\delta$ T cells points to a ubiquitous expression of such an APM, including even $\gamma\delta$ T cells, as they are able to activate themselves. Only human cells are capable of presenting the stimulatory antigens to human Vγ9/Vδ2 T cells, while murine, hamster, and some monkey cell lines are not.

The α subunit of the mitochondrial F1-ATPase was reported to be directly recognized by the Vγ9/Vδ2 TCR aided by the presence of apolipoprotein A-1 in the media and this finding was confirmed by soluble Vγ9/Vδ2 TCR binding to immobilized F1-ATPase in surface plasmon resonance analysis [Scotet et al. 2005]. However, F1-ATPase is absent from some human APC that are capable of presenting antigens to $\gamma\delta$ T cells, thus challenging the claimed role of F1-ATPase as APM for Vγ9/Vδ2 T cells.

4. Non-peptidic antigen-specific T cells

The identification of TCR $\gamma\delta$ cells recognizing phosphorylated non-peptidic metabolite antigens and of CD1-restricted TCR $\alpha\beta$ cells reacting to lipid antigens [Porcelli et al. 1996], inspired a large number of studies on non-peptide-specific T cells and promoted a new branch of modern immunology that bridges innate and adaptive immunity. Two hallmark papers described CD1-restricted TCR $\gamma\delta$ and TCR $\alpha\beta$ cells [Porcelli et al. 1989] and identified $CD4^+$ T cells in MHC class II-deficient mice [Cardell et al. 1995], which were later found to be CD1-restricted and lipid-specific. CD1-restricted T cells have initially been linked to microbial infections [Park & Bendelac 2000; Porcelli & Modlin 1999; Sieling et al. 2000] and mycobacterial lipids [Moody et al. 2000].

4.1. $\alpha\beta$ T cells

T cells restricted by group 1 CD1 molecules have been isolated and characterized. Due to lack of tools allowing direct identification of group 1 CD1-restricted T cells, the knowledge about the TCR repertoire, surface marker expression and modulation during diseases remains very limited. A common TCR $\alpha\beta$ interaction pattern with antigen complexes has only been proposed for CD1b [Melian et al. 2000].

CD1d-restricted T cells were called natural killer T (NKT) cells because of their expression of natural killer (NK) cell markers [Terabe & Berzofsky 2008]. The NK marker expression is not a distinctive feature of NKT cells as some peptide-specific classical T cells are able to upregulate these markers upon activation. Therefore, NKT cells were redefined by their CD1d-restriction [Godfrey & Kronenberg 2004 a]. Two sets of NKT cells are able to recognize CD1d and are termed type 1 and type 2 NKT cells. Type 1 NKT cells are defined by their invariant TCR Vα24-Jα18 chain in humans and Vα14-Jα18 chain in mice, as well named invariant NKT (iNKT) cells after their TCR α-chain, whereas type 2 NKT cells use a more diverse TCR repertoire without any bias in the Vα and Vβ chains.

4.1.1. iNKT cells

Type 1 NKT cells pair their invariant TCR Vα24-Jα18 mostly with TCR Vβ11 in humans and with TCR Vβ8.2, 7 or 2 in mice. The term semi-invariant arose with the finding of a diverse TCR β repertoire in mice [Matsuda et al. 2001] associated with the conserved Vα14 to Jα18 rearranged TCR α gene segments with a single N-nucleotide addition coding the third base of a characteristic glycine. The frequency of this particular TCR considering the TCR α gene segments should be around one in one million T cells. Astonishingly, iNKT cells are found in the blood of mice with a frequency of $\approx$ 1% and in human peripheral blood mononuclear cells (PBMC) of $\approx$ 0.1% and therefore three to four logs more often than expected (see [Godfrey & Berzins 2007] Table 1). In 1997, a pan-activatory ligand for iNKT cells, the glycolipid αGC, was found presented by CD1d [Kawano et al. 1997]. This ligand boosted the research on iNKT cells tremendously by offering a tool to directly identify iNKT cells by αGC-loaded CD1d tetramers, allowing pan-iNKT cell-expansion from total PBMC, clinical applications and iNKT cell-studies in mouse models, as both mouse and human iNKT cells react to this glycosphingolipid. Ten years later, when the iNKT-αGC-CD1d co-crystal was solved, it became clear how the TCR of iNKT cells interacts with the αGC-CD1d complex. The iNKT TCR docks longitudinally and slightly laterally over the F' pocket of CD1d. This is different from the diagonal binding of a conventional αβ TCR which docks on the top of peptide-MHC complexes [Borg et al. 2007]. Another peculiarity of the iNKT TCR mode of binding is the fact that the αGC-TCR contact is established solely by the TCR α-chain whereas the TCR β-chain contacts the CD1d molecule without touching αGC [Wun et al. 2008]. On the other hand, both α- and β-chain of conventional TCR make contacts with the peptide-antigen [Marrack et al. 2008; Rudolph et al. 2006; Wang & Reinherz 2002]. The fact that most of the key residues contributing to the TCR-αGC-CD1d interactions are conserved among species, explains the reported capacity of mouse CD1d to present αGC to human iNKT cells and vice versa.

CD1d and Jα18 knockout mice were essential to address iNKT cell-development. Initial investigations revealed similarities and main differences to conventional T cells. iNKT cells and conventional T cells emerge from DP thymocytes and share common precursors but the former are positively selected by CD1d-expressing DP thymocytes and the latter on MHC class I and II molecules of thymic epithelial cells [Gapin et al. 2001]. Two main models have been proposed for iNKT cell development, namely the instructional and the precommitment model claiming the TCR-dependence or -independence to become an iNKT cell (see [Kronenberg 2005] Figure 2), respectively. The selecting ligand remains controversial even though the self glycosphingolipid iGb3 has been suggested due to the decrease of 95% in the thymic NKT population in β-hexosaminidase-B-deficient mice [Bendelac et al. 2007] and the inability of β-hexosaminidase-B-deficient thymocytes to activate Vα14 NKT cell hybridomas [Zhou et al. 2004 a]. The absence of iGb3 in the thymus [Speak et al. 2007] and a normal and functional iNKT cell compartment in iGb3 synthase-deficient mice, which do not synthesize this glycolipid, [Porubsky et al. 2007] argue against selection by iGb3. Thus the selecting ligand(s) remain unknown.

iNKT cells are activated several times during their thymic development and have to pass at least two control points while progressing from a DP stage via four 'immature' stages, defined by their CD24, CD44, DX5 and NK1.1 expression pattern, till they egress to the periphery (see [Godfrey & Berzins 2007] Figure 3). The lately identified promyelocytic leukemia zinc-finger (PLZF) transcription factor [Kovalovsky et al. 2008; Savage et al. 2008] seems limited in expression among immune cells to iNKT cells and is crucial for the acquirement of peripheral iNKT characteristics (see [Gapin 2008] Figure 1). It appears at or just after positive selection of iNKT cells and is found in human iNKT cells as well where it may have equal functions as the protein is highly conserved.

Recent findings shed more light on the early stages of iNKT cell development during the $CD4^-CD8^-$ double-negative (DN) stage that can be subdivided into 4 sequential stages according to the 4 possible permutations of CD25 and CD44 expression. During progression from DN3 to DN4 stage conventional T cells rearrange their TCR β-chain that, if productive, pairs with the pre-TCR α-chain and permits progression to the DP stage. At this stage the TCR α-chain locus rearranges and allows transcription of the TCR Vα-Jα-Cα chain. In contrast, iNKT DN4 cells express Vα14-Jα18 transcripts intracellularly, but no surface TCR. These cells are similar in wild type (WT) and CD1d knockout mice and have the same potential to give rise to iNKT cells when selected on WT thymocytes [Dashtsoodol et al. 2008].

Several lines of evidence, e.g. the activated phenotype of peripheral iNKT cells as well as the lack of iNKT cells in Jα18-deficient mice (the Jα gene segments are contained in the CDR3α contacting the antigen), indicate the presence of endogenous ligands for iNKT cells. The self glycosphingolipid iGb3 has been proposed as one endogenous ligand capable of inducing positive selection of iNKT cells. However, subsequent studies have shown that mice without this lipid do not have alterations in iNKT cells, thus suggesting that other ligands are responsible for positive selection [Godfrey et al. 2004 b]. Several endogenous lipids may bind to CD1d [Giabbai et al. 2005; Zajonc et al. 2005 a], but their involvement in positive selection of iNKT cells has not been proven.

Sulfatide, not activatory for human iNKT cells [Franchini et al. 2007], was observed to drive human CD1d-restricted T cells to proliferate and release cytokines (our unpublished results). The caveolar-enriched ganglioside monosialotetrahexosylganglioside (GM1) is another self lipid able to bind to CD1d and compete with αGC (our unpublished observations).

Recent studies were trying to better describe CD1d-bound self-ligands using mass spectrometry [Cox et al. 2009; Fox et al. 2009; Yuan et al. 2009]. These investigations showed that lysophospholipids are stimulatory for human iNKT cells [Fox et al. 2009]. Even though a large quantity of lipids, amongst them several linked to immune diseases, are associated with CD1d [Cox et al. 2009], solely mono-acyl-lysophosphatidylcholine was antigenic towards iNKT cell clones and total iNKT cells [Fox et al. 2009]. During trafficking through the cell, CD1d is loaded with several ligands [Yuan et al. 2009], namely within the ER mainly with phosphatidylcholine, the most abundant eukaryotic phospholipid; within LE with phosphatidylcholine, sphingomyelin, and lysophospholipids; and possibly within the Golgi with sphingomyelin. Interestingly, autoreactive human iNKT cells recognize CD1d molecules loaded with endogenous ligands even when CD1d is engineered in order to avoid lysosomal trafficking, whereas mouse autoreactive iNKT cells require lysosomal routing of CD1d [Chen et al. 2007; Chiu et al. 1999; Chiu et al. 2002]. Whether this is associated with different antigen specificity of human and mouse autoreactive iNKT cells [De Libero & Mori 2009 a] remains to be investigated. The identification of the endogenous stimulatory ligands [De Libero et al. 2009 b] is of major relevance as iNKT cells are important regulatory T cells [Godfrey & Kronenberg 2004 a] and are also involved in autoimmunity and cancer cell recognition and activation of innate immune cells [Cohen et al. 2009; Taniguchi et al. 2010].

4.1.2. NKT cells

Early studies in mice claimed a tissue-specific segregation and development of CD1d-dependent and -independent NKT cells [Eberl et al. 1999]. The definition of NKT cells in this report is in contradiction with the one used later defining all NKT cells as CD1d-dependent. The cells described in that study could indeed be a diverse, NK marker-positive T cell population developing extrathymically with a unknown restriction or maybe restricted by CD1d2, that is not sufficient to drive NKT development [Chen et al. 1999] but could be involved in several disease outcomes [Qian et al. 2010; Sakai et al. 2002].

A rare human CD8 memory NK marker-positive T cell-population, that was not CD1d-restricted and possibly of extrathymic origin, was described to recruit innate immune cells upon Fas engagement and to efficiently kill target cells [Giroux & Denis 2005].

In humans, TCR Vα24⁻ NKT cells that recognized αGC in the context of CD1d were found to be either CD8αβ- or CD4-positive [Gadola et al. 2002 b]. This could be explained by epitope mimicry of αGC for certain NKT cells or be repugnant to the definition of iNKT cells but supports the instructive developmental model of TCR-antigen driven selection. The TCR Vα24⁻ cell population staining bright with αGC-loaded CD1d tetramers is TCR Vβ11⁺ whereas the duller population shows a diverse TCR Vβ usage. This is in line with a model of one iNKT

TCR crystal and two αGC-CD1d–specific Vβ11$^+$ TCR crystals that show a similar mode of recognition by all three TCR [Gadola et al. 2006 b].

Studies conducted in CD1d knockout and Jα18 knockout mice in some cases revealed different susceptibilities to infection, thus pointing to functionally distinct NKT and iNKT cell populations, respectively [Tupin et al. 2007].

4.2. γδ T cells

γδ and αβ T cells arise from a common precursor in the thymus, but diverge into separate lineages early in ontogeny [Hayes & Love 2007]. Differentiation along the γδ lineage likewise starts at the immature DN stage, but lacks the pre-TCR, DP, and CD4/CD8 co-receptor single-positive stages as well as a developmental extensive proliferation [Xiong & Raulet 2007]. Instead, most γδ lineage cells remain DN, proliferate barely or not at all, and express the mature TCR γδ complex [Taghon & Rothenberg 2008]. Still, signals transduced through the TCR are essential in maintaining thymic γδ T cell differentiation and maturation [Hayday 2000].

Human TCR γδ cells use a limited repertoire of Vγ and Vδ with junctional diversity in their chains [Carding & Egan 2002; Chien & Bonneville 2006], namely one of three major Vδ gene segments (Vδ1, Vδ2, or Vδ3) and one of 6 different Vγ gene segments. As in mice, the combination of Vγ and Vδ segments in the TCR γδ heterodimer determines their tissue distribution and antigen recognition pattern. The majority (50-80%) of human γδ T cells express a TCR heterodimer composed of the Vγ9 and Vδ2 chains. Vγ9/Vδ2 T cells are unique to humans and primates and represent 0.5 to 5% of the lymphocyte population among PBMC and are found in tonsils and in spleen [Casorati et al. 1989], too. Intestinal TCR γδ cells preferentially pair Vδ1 and Vδ3 chains with members of the Vγ1 gene family [Carding & Egan 2002; De Libero et al. 1993].

Initially, TCR Vγ9/Vδ2 cells were characterized by their non-specific cytotoxicity against lymphomas and erythroleukemia cells that was later shown to depend on engagement of the TCR [Bukowski et al. 1995]. Around the same time, the order of potency of several ligands stimulating diverse TCR Vγ9/Vδ2 cell clones was found constant [Buerk et al. 1995], suggesting a type of antigen recognition different from that of TCR αβ cells interacting with peptide-MHC complexes. This pan-activating potential and the fact that peripheral TCR Vγ9/Vδ2 cells present a memory-like state responding immediately to antigenic challenges without apparent priming [Tough & Sprent 1998] are still considered indirect evidences that γδ T cells are close to innate immune cells.

A series of investigations have confirmed that all tested Vγ9/Vδ2 T cells recognize (E)-4-hydroxy-3-methyl-but-2-enyl pyrophosphate (HMBPP), a small microbial compound that is a natural intermediate of the MEP pathway of IPP biosynthesis in microbes [Eberl et al. 2003]. HMBPP is an essential metabolite in most pathogenic bacteria including *Mycobacterium tuberculosis* and malaria parasites. Certain bacterial species, such as *Streptococcus*, *Staphylococcus*, and *Borrelia*, lack the MEP pathway and utilize the mevalonate pathway for IPP synthesis instead. The Vγ9/Vδ2 TCR crystal structure confirmed the presence of a basic, positively charged region in the binding groove that could directly interact with the negatively charged pyrophosphate moiety of the antigen [Allison et al. 2001]. This charge is essential for antigen recognition. Indeed, site-directed mutagenesis of the positively charged lysine in the Jγ1.2 segment of the CDR3γ to a negatively charged glutamate abrogated Vγ9/Vδ2 T cell stimulation by prenyl pyrophosphates as well as aminobisphosphonates and alkylamines, demonstrating that this basic residue is necessary for recognition of all non-peptidic antigens [Miyagawa et al. 2001].

Vγ9/Vδ2 T cells are assumed to sense 'danger' signals of invading pathogens (e.g. in tuberculosis, salmonellosis, brucellosis, listeriosis, and malaria). This attributes $\gamma\delta$ T cells an important role in immunity to pathogens [Morita et al. 2000]. In addition, $\gamma\delta$ T cells appear to have a sentinel function by reacting to altered self as for instance against tumors. Recent studies have shown that recognition of both infected cells and tumor cells is associated with changes in the ubiquitous mevalonate pathway and with increased cellular levels of IPP [Gober et al. 2003; Kistowska et al. 2008]. Therefore, the system has evolved to sense subtle modifications of the vital mevalonate metabolic pathway.

$\gamma\delta$ T cells may also contribute to the integrity of the epithelium during infection or injury facilitating tissue healing [Jameson et al. 2002], thereby limiting the spread of pathogens within the host [Havran et al. 1991]. In $\gamma\delta$ T cells, TCR-mediated activation is complemented by activating and inhibitory NK receptors, which modulate the response to different antigen doses [Carena et al. 1997].

Unlike most $\alpha\beta$ T cells, only a marginal fraction of circulating Vγ9/Vδ2 T cells express CCR7, preventing them from entering lymph nodes. Instead, they express unique subsets of inflammatory chemokine receptors which allow homing to sites of inflammation. $\gamma\delta$ T cells rapidly upregulate CCR7 and B-cell costimulatory ligands (CD40L, OX40, CD70, and ICOS) upon activation and help B cells to traffic to lymph nodes and to mature [Morita et al. 2007].

Similar to type 2 iNKT cells there exist non-Vδ2 T cells and alike they possess extensive junctional diversity in their Vδ1 and Vδ3 TCR chains. Non-Vδ2 T cells are mostly localized inside epithelial-rich (mucosal, intestinal and skin) tissues [Hayday & Tigelaar 2003]. Two cases of Vδ1 T cells have been reported to be CD1d- [Russano et al. 2006] or CD1c-restricted [Spada et al. 2000] under special circumstances such as during allergy or after expansion with mycobacterial extracts, respectively. Rare Vδ1 clones reactive to MHC, MHC-like [Groh et al. 1998], or non-MHC molecules were established [Carding & Egan 2002; Thedrez et al. 2007] but neither cognate interaction by surface plasmon resonance studies nor blocking of Vδ1 T cell-responses by specific antibodies has been reported. These experiments suggest recognition of a new class of conserved stress-induced antigens by non-Vδ2 T cells, which expand during infections with intra- and extracellular bacteria as well as viruses. In most instances, Vδ1 expansion is not triggered by stimuli derived from pathogens but instead follows recognition of endogenous gene products presumably upregulated during infection. The endogenous antigens recognized by non-Vδ2 T cells have not been characterized yet.

Despite recent advances in the $\gamma\delta$ T cell field [Kronenberg & Havran 2007], many aspects of $\gamma\delta$ T cells and of their functions remain unknown.

Part II.

Materials and Methods

5. Materials

All external links given as URL were checked and working at the time the thesis was written. The author of this thesis neither has influence on nor can be made responsible for potential moving of the source location and for changes in content of the linked source.

Further information and hints about current protocols can be found in the 2nd edition of "T cell protocols" edited by G. De Libero (http://dx.doi.org/10.1007/978-1-60327-527-9).

5.1. Antibodies

Whenever the isotype of an antibody is known it was listed as immunoglobulin (Ig) plus its class and subclass.

Most antibodies were used for different applications as per manufacturer's instructions and tested in further experimental designs according to study needs. Most used procedures include FACS (see section 7.8, p.47), CLSM (see section 7.6, p.46) and ELISA (see section 7.7, p.46). For every type of experiment all used antibodies were carefully titrated to give optimal results and least background.

Table 5.1   Antibodies, human

Specificity	Clone	Host	Isotype	Reference
β_2m	BBM.1	mouse	IgG2b	[Brodsky et al. 1979]
BirA peptide	BIR1.4	mouse	IgG2b	
CD1a	OKT6	mouse	IgG1	
CD1b	BCD1b3.1	mouse	IgG1, κ	
CD1c	F10/21A3	mouse	IgG1	
CD1d	42	mouse	IgG1	
CD1d	51	mouse	IgG2b	
CD1e	1-22	mouse	IgG1	
CD1e	20.6	mouse	IgG1	

continued on next page

5. Materials

continued from previous page

Specificity	Clone	Host	Isotype	Reference
CD3	OKT3	mouse	IgG2a	
CD3	S4.1 or 7D6	mouse	IgG2a, κ	
CD3	TR66	mouse	IgG1	
CD3	UCHT1	mouse	IgG1	
CD4	MT310	mouse	IgG1, κ	
CD4	RPA-T4	mouse	IgG1, κ	
CD4	S3.5	mouse	IgG2a	
CD8	3B5	mouse	IgG2a	
CD8	DK25	mouse	IgG1, κ	
CD8	LT8	mouse	IgG1	
CD25	BC96	mouse	IgG1, κ	
CD25	M-A251	mouse	IgG1, κ	
CD27	CLB-27/1	mouse	IgG2a	
CD27	LT-27	mouse	IgG2a	
CD27	M-T271	mouse	IgG1, κ	
CD39	A1	mouse	IgG1	
CD45	J33	mouse	IgG1	
CD45RA	L48	mouse	IgG1, κ	
CD45RA	MEM56	mouse	IgG2b	
CD69	CH/4	mouse	IgG2a	
CD69	L78	mouse	IgG1	
CD71 (TfR)				
CD74 (Ii)	LN-2 or B318	mouse	IgG1	
CD94		mouse		
GM-CSF (capture)	BVD2-23B6	rat	IgG2a	
GM-CSF (detection)	BVD2-21C11	rat	IgG2a	
GM-CSF (capture)	6804	mouse	IgG1	
GM-CSF (detection)	3209	mouse	IgG1	
HLA DP	B7/21.2	mouse	IgG1, κ	
HLA DQβ	XIV 466.2	mouse		
HLA DQ1	BT3/4	mouse		
HLA DQ2	XIII 358.4	mouse		
HLA DQw1	Genox 3.53	mouse	IgG1	[Brodsky et al. 1980]
HLA DR	2/72	mouse		
HLA DR	D1.12	mouse	IgG2a	
HLA DR	L227	mouse	IgG1, κ	[Lampson & Levy 1980]

continued on next page

continued from previous page

Specificity	Clone	Host	Isotype	Reference
HLA DR	L243	mouse	IgG2a, κ	[Lampson & Levy 1980]
HLA DRα	DA6.147	mouse	IgG1	[Guy et al. 1982]
HLA DR/DP/DQ	IVA12	mouse	IgG1, κ	[Shaw et al. 1985]
HLA DR/DQ	9.3F10	mouse	IgG2a	[Van Voorhis et al. 1983]
IFN-γ (capture)	HB8700	mouse	IgG1	
IFN-γ (detection)	γ69-2GV	mouse	IgG1	[Gallati et al. 1987]
IL-4 (capture)	8D4-8	mouse	IgG1	
IL-4 (detection)	MP4-25D2	rat	IgG1	
IL-4 (detection)	3H4			
IL-4 (detection)	8F12	mouse	IgG1	
IL-8 (capture)	JK8-1	mouse	IgG1	
IL-8 (detection)	JK8-2	mouse	IgG1	
IL-17 (capture)	41809	mouse	IgG2b	
IL-17 (detection)		goat	polyclonal IgG	
LAG-3	17B4	mouse	IgG1	
MHC class I	W6/32	mouse	IgG2a, κ	
MRP5	C-17	goat	polyclonal	
MRP5	M5I-1	rat	IgG2a	
MRP5	P-20	goat	polyclonal	
NKG2D	ON72	mouse	IgG1	
NKp30	Z25	mouse	IgG1	
NKp44	Z231	mouse	IgG1	
NKp46	BAB281	mouse	IgG1	
TCR Vα24	C15	mouse	IgG1, κ	
TCR Vβ11	C21	mouse	IgG1, κ	[Dellabona et al. 1994]
TCR Vγ9	B3	mouse	IgG1	
TCR Vδ2	4G6	mouse	IgG1	
TNF-α (capture)	MAb1	mouse	IgG1	
TNF-α (detection)	MAb11	mouse	IgG1	
TNF-α (capture)	357-101-4	mouse	IgG1	
TNF-α (detection)	2-179-E11	mouse	IgG1	

5. Materials

Table 5.2 Antibodies, mouse

Specificity	Clone	Host	Isotype	Reference
IFN-γ (capture)	R4-6A2	rat	IgG1	
IFN-γ (detection)	XMG1.2	rat	IgG1	
IL-2 (capture)	JES6-1A12	rat	IgG2a	
IL-2 (detection)	JES6-5H4	rat	IgG2b	
IL-4 (capture)	11B11	rat	IgG1	
IL-4 (detection)	BVD6-24G2	rat	IgG1	

5.2. Cell lines and clones

Table 5.3 Cell lines, human

Name	Type	Reference
A-375	malignant melanoma	[Giard et al. 1973]
C1R	B lymphoblastoma	[Storkus et al. 1989]
HeLa	negroid cervix epitheloid carcinoma	[Scherer & Hoogasian 1954]
HL-60	caucasian promyelocytic leukemia	[Collins et al. 1977]
HMEC-1	dermal microvascular endothelial cell line	[Ades et al. 1992]
K-562	caucasian chronic myelogenous leukemia	[Lozzio & Lozzio 1975]
Molt-4	acute T lymphoblastic leukemia	[Minowada et al. 1972]
T2	lymphoblastoma	[Salter et al. 1985]
THP-1	monocytic leukemia	[Tsuchiya et al. 1980]

Table 5.4 Cell lines, mouse

Name	Type	Reference
CTLL-2	IL-2 dependent C57BL/6 T lymphoblast	[Gillis & Smith 1977]
J558	BALB/c plasmacytoma	[Weigert et al. 1970]
P3X63Ag8	BALB/c plasmacytoma	[Koehler & Milstein 1975]
P3X63Ag8.653	BALB/c plasmacytoma	[Kearney et al. 1979]

All cell lines in the laboratory were tested upon receipt and at regular intervals thereafter in accordance with institutional guidelines (see appendix SOP mycoplasma, p.199).

Table 5.5 Clones, human

Name	Type	Restriction	Specificity	Reference
BG clones	iNKT	CD1d		[Franchini et al. 2007]
DS1B9c	$\alpha\beta$T	CD1c	Sulfatide	[Shamshiev et al. 2002]
DS1C9b	$\alpha\beta$T	CD1b	Sulfatide	[Shamshiev et al. 2002]
DS2C13a	$\alpha\beta$T	CD1a	Sulfatide	[Shamshiev et al. 2002]
G2B9	$\gamma\delta$T		HMBPP, IPP	[Casorati et al. 1989]
GG33a	$\alpha\beta$T	CD1b	GM1	[Shamshiev et al. 1999]
JS7	iNKT	CD1d		[Xia et al. 2006]
JS63	iNKT	CD1d		[Xia et al. 2006]
K34A6.2	$\alpha\beta$T	CD1a	Sulfatide	[Shamshiev et al. 2002]
K34B9.1	$\alpha\beta$T	CD1a	Sulfatide	[Shamshiev et al. 2002]
S17d	$\alpha\beta$T	CD1d	Sulfatide	
S33d	$\alpha\beta$T	CD1d	Sulfatide	
S38d	$\alpha\beta$T	CD1d	Sulfatide	
SaGC lines	iNKT	CD1d		[Kyriakakis et al. 2009]
VM-D5	iNKT	CD1d		

5. Materials

Table 5.6 Hybridomas, mouse

Name	Type	Restriction	Specificity	Reference
FF4	iNKT	CD1d		[Schuemann et al. 2007]
FF5	iNKT	CD1d		[Schuemann et al. 2007]
FF13	iNKT	CD1d		[Schuemann et al. 2007]

All clones and hybridomas mentioned above were established in the laboratory. These cells were tested after establishment and periodically thereafter in accordance with institutional guidelines (see appendix SOP mycoplasma, p.199).

5.3. Compounds

Table 5.7 Compounds

Name	Class	APM [i]	Responder	Reference
α-galactosylceramide (**KRN7000**)	GSL [ii]	CD1d	iNKT	[Kobayashi et al. 1995]
α-**sulfatide**	GSL	CD1d	iNKT	[Franchini et al. 2007]
7-aza- and 7-thiasphingosines	SL [iii]	CD1d		[Mathew et al. 2009 c]
C-analogue of sulfatide	GSL	CD1a		[Modica et al. 2006]
KRN7000 S-glycoside analogue	GSL	CD1d		[Rajan et al. 2009]
Sphingolipid analogues	SL	CD1d		[Mathew et al. 2009 b]

[i] Antigen-presenting molecule(s)
[ii] Glycosphingolipid
[iii] Sphingolipid

The compounds in (see Table 5.7, p.36) were synthesized and/or purified by collaborators and generously provided for mutual studies or projects.

5.4. Cytokines and chemokines

Human GM-CSF is produced from J558 cells transfected with the human GM-CSF gene (clone 36.17, pMCK-gpt-hGMCSF) in FCSM (see section 5.6.1, p.39) with 50 µM β-mercaptoethanol after selection in FCSM with 50 µM β-mercaptoethanol plus 5 µg/ml mycophenolic acid and 125 µg/ml xanthine and tested in a classical sandwich ELISA (see section 7.7, p.46) against commercial recombinant human GM-CSF standard.

The most prominent of the interleukins (IL) affecting T cells, human IL-2 is produced from J558 cells transfected with the human IL-2 gene (clone 2, pMCK-gpt-hIL2) in FCSM with 50 µM β-mercaptoethanol after selection in FCSM with 50 µM β-mercaptoethanol plus 5 µg/ml mycophenolic acid and 125 µg/ml xanthine and tested on CTLL-2 cells against commercial recombinant human IL-2 standard.

Human IL-4 is produced from P3X63Ag8.653 cells transfected with the human IL-4 gene in FCSM with 50 µM β-mercaptoethanol after selection in FCSM with 50 µM β-mercaptoethanol plus 5 µg/ml mycophenolic acid and 125 µg/ml xanthine and tested against commercial recombinant human IL-4 standard in a sandwich ELISA.

Human IL-6 is produced from P3X63Ag8.653 cells transfected with the human IL-6 gene (clone X6310-46, BCMGSneo-hIL-6) in FCSM with 50 µM β-mercaptoethanol after selection in FCSM with 50 µM β-mercaptoethanol plus 5 µg/ml mycophenolic acid and 125 µg/ml xanthine and tested against commercial recombinant human IL-6 standard in ELISA.

All cytokine-containing media were filtered through Stericup GP Express Plus 0.45 µm membrane filters (Millipore, Zug, Switzerland), tested, aliquotted and stored at -20 °C.

Human IL-8 (BioLegend, San Diego, California, USA), IL-17 (R&D Systems, Abingdon, United Kingdom), IFN-γ (BD Biosciences, Allschwil, Switzerland or PeproTech, London, United Kingdom or Bender MedSystems, Vienna, Austria) and TNF-α (PeproTech or ImmunoKontact, AMS Biotechnology, Bioggio Lugano, Switzerland) were bought from commercial suppliers.

All commercial cytokines were tested, stocked at the needed concentrations (e.g. 1 µg/ml for ELISA) and stored at -70 °C.

5.5. DIMER

For preparation of soluble recombinant human CD1d, a strategy similar to the one introduced to obtain soluble recombinant human CD1b [Garcia-Alles et al. 2006] was developed. The cloning was adapted from published methods [Manolova et al. 2003, 2006]. Shortly, human CD1d cDNA, devoid of the transmembrane and cytoplasmic coding sequence, was amplified from C1R-hCD1d cells [Brossay et al. 1998], that express the full-length protein, using the 5'hCD1d-XhoI and 3'hCD1d-P-ClaI primer pair and the Advantage® HF 2 DNA polymerase (Clontech, Takara Bio Europe, Saint-Germain-en-Laye, France).

Human CD1D amplification primers

5' - CTC GAG ATA TGG GGT GCC TGC TG - 3' (5'hCD1d-XhoI)

5' - TTA TCG ATA GGC CCA CCC CAG TAG AGG AC - 3' (3'hCD1d-P-ClaI)

The product was cloned into pCR2.1-TOPO (Invitrogen, Basel, Switzerland) and sequenced. The XhoI/ClaI fragment was ligated into pBluescript-Bir [Nowbakht et al. 2005], a pBluescript II KS+ (Stratagene, Agilent Technologies, Basel, Switzerland) based vector containing a fragment encoding the #85 BirA peptide tag (GGGLNDIFEAQKIEWHE) [Schatz 1993]. The soluble human CD1d-birA fusion product was excised by XhoI/NotI digestion and subcloned into the eukaryotic cDNA expression vector BCMGSneo [Karasuyama et al. 1990]. The final construct was

5. Materials

super-transfected by electroporation into J558 cells expressing human β_2m with a hygromycin resistance gene [Nowbakht et al. 2005]. After double-selection with hygromycin B and G418 (both Calbiochem, Merck, Nottingham, United Kingdom), cells were cloned, picked according to expression, expanded and frozen. The amino acid sequence of soluble CD1d is as follows.

Soluble recombinant birA-tagged human CD1d including the leader sequence [i]

[MGCLLFLLLW ALLQAWGSA][ii]E VPQRLFPLRC LQISSFANSS WTRTDGLAWL GELQTHSWSN DSDTVRSLKP WSQGTFSDQQ WETLQHIFRV YRSSFTRDVK EFAKMLRLSY PLELQVSAGC EVHPGNASNN FFHVAFQGKD ILSFQGTSWE PTQEAPLWVN LAIQVLNQDK WTRETVQWLL NGTCPQFVSG LLESGKSELK KQVKPKAWLS RGPSPGPGRL LLVCHVSGFY PKPVWVKWMR GEQEQQGTQP GDILPNADET WYLRATLDVV AGEAAGLSCR VKHSSLEGQD IVLYWGG[PID KL][iii][GGGLNDIF EAQKIEWHE][iv]

For every production of soluble recombinant human CD1d, cells were double-selected with hygromycin B and G418 for two weeks, adapted to and then cultured under low serum growth conditions (see subsection 5.6.1, p.39). Soluble birA-tagged CD1d was purified similarly to reported protocols [Garcia-Alles et al. 2006]. Briefly, harvested supernatants were precipitated in 40% ammonium sulfate and dialyzed against phosphate buffered saline (PBS). Soluble human CD1d was purified from total proteins by isoelectric focusing (IEF) in a Rotofor™ preparative IEF cell (Bio-Rad, Reinach, Switzerland) with 1.5% Bio-Lyte 5/8 Ampholyte (Bio-Rad).

Table 5.8   Soluble recombinant birA-tagged human CD1 proteins

protein	Uniprot	amino acids	molecular weight	isoelectric point [v]
CD1a	P06126	302	34.6 kDa	5.87
CD1b	P29016	302	33.5 kDa	5.21
CD1c	P29017	302	34.1 kDa	5.55
CD1d	P15813	300	34.0 kDa	6.21
BSA	P02769	607	69.3 kDa	5.82

[v] calculations by http://www.expasy.ch/tools/pi_tool.html (with birA peptides; without leader peptides, β_2m) and glycosylation

IEF fractions were tested by Western blotting with the BIR1.4 monoclonal antibody (mAb) and pooled. The pooled fractions were dialyzed against PBS, stocked and stored at -70 °C until use. IEF was primarily performed to eliminate bovine serum albumin (BSA) and other proteins of no interest contained in cell culture supernatants.

For preparation of the DIMER, IEF-purified soluble recombinant birA-tagged CD1d was loaded with molar excess of the antigen of interest for 45 min at room temperature (RT) before addition of twofold molar excess of BIR1.4 mAb for another 15 min at RT. DIMER were used immediately or stored at 4 °C. They remained stable for more than one year after preparation.

[i]   319 amino acids including leader and BirA target peptides
[ii] **Amino acids of the leader peptide are given bold**
[iii] Amino acids added for in-frame cloning are given bold and blue
[iv] Biotinylation substrate peptide consensus sequence is given bold and red

5.6. Media and buffers

5.6.1. Media basis

All media were prepared in RPMI 1640 (Gibco, Invitrogen, Basel, Switzerland) with 100 µM non-essential amino acids, 1 mM sodium pyruvate, 100 µg/ml kanamycin, 2 mM GlutaMAX^{TM}-I (all Gibco) and filtered through Stericup GP Express Plus 0.22 µm membrane filters (Millipore).

Fetal calf serum medium

The fetal calf serum medium (FCSM) was prepared by addition of 10% fetal calf serum (Gibco) to the basic medium.

Human serum medium

The human serum medium (HSM) is composed of basic medium with 100 U/ml human IL-2 and 5% AB-positive human serum (Blutspendezentrum SRK beider Basel, Basel, Switzerland).

5.6.2. Buffers

ELISA blocking solution

Released cytokines were measured by ELISA in sterile filtered PBS plus 10 mg/ml bovine serum albumin (PAA, Cölbe, Germany or Sigma-Aldrich, Basel, Switzerland) and 0.05% Tween-20 (Sigma-Aldrich).

ELISA washing solution

The ELISA plates were washed with PBS and 0.05% Tween-20. The ELISA washing solution was prepared as 10x stock from PBS powder (AppliChem, Darmstadt, Germany) and diluted in MilliQ water before use.

Erythrocyte lyzing buffer

Erythrocyte lysis was done in PBS containing 174 mM ammonium chloride (NH_4Cl), 10 mM potassium bicarbonate ($KHCO_3$), 0.1 mM Na_2EDTA, and 10 mM sodium azide (NaN_3).

FACS buffer

Surface staining of cells was performed in PBS supplemented with 0.5% human albumin (kind gift of the Swiss Red Cross) and 0.02% sodium azide (NaN_3).

MACS buffer

Isolation of cell populations by magnetic-activated cell sorting (MACS) used a sterile filtered buffer composed of 2 mM ethylenediaminetetraacetate (EDTA) and 0.5% bovine serum albumin in PBS.

Saponin buffer

Intracellular staining of fixed cells was achieved by permeabilizing the cells in PBS containing 0.5% (v/v) human albumin and 0.02% (v/v) sodium azide (NaN_3) plus 0.1% (w/v) saponin (from Quillaja bark, Sigma-Aldrich).

5.7. Primers

5.7.1. Mycoplasma

A standard operating protocol (SOP) for detection of mycoplasma has been developed (see appendix SOP mycoplasma, p.199) for the Department of Biomedicine, University Hospital Basel.

6. Cell culture techniques

6.1. Cloning, cells

Isolation of iNKT cell clones from PBMC of healthy donors [Franchini et al. 2007] was described before. The general procedure for cloning T cells from PBMC is as follows. Use fresh HSM and 5×10^5 irradiated feeders per ml plus 1 µg/ml phytohemagglutinin final. Instead of phytohemagglutinin T cell specific antigens can be used. In the case of cloning with a specific antigen the amount of human IL-2 should be reduced to suboptimal doses e.g. 5 to 10 U/ml until cells start blasting. The amounts of medium are calculated for one threefold dilution series to plate 1 Terasaki plate (Nunc, Thermo Fisher Scientific, Roskilde, Denmark) with 10 cells per well (10 cells in 20 µl matching 500 cells per ml, e.g. 4 to 12 ml) and 2 Terasaki plates with 3 cells per well (3 cells in 20 µl matching 150 cells per ml, e.g. 4 to 9 ml) and 4 Terasaki plates with 1 cell per well (1 cell in 20 µl matching 50 cells per ml, e.g. 4 to 9 ml) and 10 Terasaki plates with 0.3 cell per well (0.3 cell in 20 µl matching 15 cells per ml, e.g. 4 to 9 ml). Dilution series are plated with a special cloning tool and plates are placed in a cell culture incubator (see section 6.6, p.43) as soon as possible to avoid medium evaporation and thereby drying out of the cells. The cloning is left untouched for 3 to 4 d before checking for contamination. Then the cloning is kept unattended until day 12 to 14 before growing clones are counted and thereof the plating efficiency is computed. Expanding clones are picked from the lowest starting cell per well concentration upwards and transferred to 96-well plates. Once 2 96-wells are full, screening for desired clones is initiated. All clones should be named as follows. The first or two leading letters indicate the source or donor, the next letter is given by the antigen used if any, the following letter is assigned according to the plating dilution of the clone (A meaning 0.3 cell per well, B meaning 1 cell per well, C meaning 3 cells per well, and D meaning 10 cells per well), and thereupon clones are numbered increasingly.

6.2. Cloning, lines

Use restimulation protocol to expand lines from different sources. Phytohemagglutinin may be changed for T cell specific antigens at will. Antigen-specific cell expansion should be conducted under suboptimal doses of human IL-2 (5 to 10 U/ml) for the first days until blasts become visible. Thereafter human IL-2 should be added at the optimal dose of 100 U/ml as the advantage of the blasts will not be overtaken by other human IL-2 receptor expressing cells anymore.

6.3. Cytospin

Cells were collected and washed twice in cold PBS. Fixation of cells with paraformaldehyde (PFA) was performed as optional step in some cases. Cells were diluted to 0.5 to 1×10^6 per ml to reach 1 to 2×10^5 cells per spot of 200 µl. Cells were spun on appropriate glass slides (gelatine/chromalaun- or polylysine-coated) for 8 min at 800 revolutions per minute (rpm) (134 g) with fast acceleration. Cell spots were air dried completely. Acetone fixation of cells was performed as optional step in some cases. Cytospins were frozen at -70 °C until use. After thawing slides were dried at 25 °C completely (only if they were not fixed yet) before proceeding to re-naturing, possible fixation, and staining.

6.4. Depletion and enrichment

6.4.1. Isolation of lymphocytes from tissue

Collagenase-assisted release of lymphocytes from fresh tissue

Collagenase-assisted release of lymphocytes from fresh tissue was performed in accordance with previously established protocols [De Libero et al. 1993] using some modifications. Tissue was kept in PBS at RT until initiation of cell isolation procedures. The biopsy was transferred to a Petri dish containing cold PBS, cells were gently scraped off the tissue surface with a scalpel and retained. The remaining tissue was cut to tiny pieces and, after extensive washing, transferred to a conical tube for digestion. The digestion to access tissue penetrating lymphocytes was performed in HSM (see section 5.6.1, p.39) plus 20 µg/ml gentamycin (Gibco), 20 µg/ml ciproxin (Bayer, Zürich, Switzerland), and 2.5 µg/ml fungizone (Gibco) including 400 U/ml collagenase type IV (Sigma-Aldrich) and 500 U/ml benzonase nuclease or 2 mg/ml deoxyribonuclease type I (Sigma-Aldrich). Scraped and collagenase-freed cells were pooled and separated from debris by Ficoll density gradient centrifugation if desired. Release of lymphocytes from fresh arterial tissue obtained during thrombo-endarterectomy was performed as described above. Isolated arterial-resident lymphocytes were maintained in HSM (see section 5.6.1, p.39).

Expansion of tissue-isolated lymphocytes

Tissue cell isolates were subjected to two rounds of stimulation with dendritic cells obtained as described [Shamshiev et al. 2000] plus the desired antigen - for iNKT cells 100 ng/ml αGC (kind gift of Kirin Breweries, Tokyo, Japan) was used as first choice - in the presence of anti-MHC class I and anti-MHC class II mAb (W6/32 and L243, both from ATCC) to avoid activation of MHC-restricted alloreactive T cells. The expanded plaque tissue cells were assessed for the presence of the cells of interest by multicolor flow cytometry (see section 7.8, p.47) and antigen presentation assays.

6.4.2. Isolation of peripheral blood mononuclear cells (PBMC)

A SOP for isolation of PBMC (see appendix SOP PBMC, p.195) was developed for our collaborators at the Cantonal Hospital Luzern. Importantly, all components were handled and complete isolation was performed at RT to assure optimal density of Ficoll. Shortly, for 1 buffy coat being approximately 50 ml tenfold concentrated blood, five 50 ml conical tubes were filled with 15 ml Ficoll (Lymphoprep, Axon Lab, Baden-Dättwil, Switzerland) each and equilibrated for a few minutes. Three times diluted blood (in PBS) was cautiously layered on top of the Ficoll. Tubes were centrifuged 20 min at 25 °C and 2000 rpm (836 g) without brake. White PBMC-rings were collected to 5 new 50 ml conical tube. PBMC were washed in PBS, to eliminate part of the platelets and thrombocytes, and pooled to 1 conical tube. Cells were counted using a hundredfold dilution. A recovery of 2×10^8 to 1×10^9 of PBMC from 1 buffy coat is expected.

CD14-positive monocytes from PBMC

PBMC were labeled with CD14 MicroBeads and passed over a MACS cell separation column or through the autoMACS Separator (all components of Miltenyi, Bergisch Gladbach, Germany) according to the manufacturer's instructions.

Monocytes can be gained, albeit less pure, by adhesion enrichment from PBMC. Therefore PBMC were plated in flasks and after 2 to 3 h non-adherent cells were discarded and adherent cells were washed carefully before further usage.

Dendritic cells (DC) from PBMC The medium used to generate dendritic cells (DC) was FCSM with addition of 10% (50 ng/ml) human GM-CSF and 8% (1000 U/ml) human IL-4. DC differentiation was checked after three to five days by surface loss of CD14 and increase or gain of antigen-presenting molecules (e.g. CD1 and MHC) as well as costimulatory molecules (e.g. CD80 and CD86).

Maturation of DC DC were matured by culturing in DC medium (see section 6.4.2, p.43) plus addition of 1 µg/ml LPS (*E.coli*, Calbiochem) and 1 ng/ml recombinant human TNF-α (PeproTech) for one to two days.

6.5. Fixation

Glutaraldehyde fixation

Cells were washed in PBS to get rid of serum proteins before fixation. Cells were resuspended in 2 to 4 ml with no more than 3×10^6 cells per ml in a 15 ml conical tube. For larger volumes a 50 ml conical tube was used. During cell washes 0.1% glutaraldehyde (Sigma-Aldrich) in PBS was prepared. Glutaraldehyde was handled under nitrogen (N_2) and in the dark whenever possible. Resuspended cells were being gently vortexed while pouring 1 volume of 0.1% glutaraldehyde to 1 volume of cells. After no more than 20 s of glutaraldehyde fixation 4 ml 200 mM L-lysine were added immediately. The solution was continuously vortexed until its color appeared yellow. Cells were washed repeatedly in 10% FCSM. Cells were re-counted to control for death due to the fixation procedure.

Paraformaldehyde (PFA) fixation

To diminish serum proteins, cells were washed in PBS. Cells were resuspended at 2×10^6 cells per ml roughly. During cell washes the stock of 4% paraformaldehyde (PFA) (Merck) in PBS was thawed. 4% PFA stock was prepared from powder as follows. PBS was alkalified with one one-hundredth volume 5 N sodium hydroxide (NaOH) to allow solubilization of PFA, PFA powder was added and mixed until solution appeared clear. The PFA solution was acidified with 5 N hydrogen chloride (HCl) until pH reaches 7-7.5 on pH-stripes. Aliquotted stocks were stored at -20 °C. While pouring 1 volume of 4% PFA to 1 volume of cells, resuspended cells were gently vortexed. PFA-fixing cells were incubated 20 min at RT with occasional shaking to avoid settling and clumping of cells. Fixation was stopped by addition of 200 mM L-lysine. Cells were washed in 10% FCSM and re-counted to exclude dead cells occurring from fixation.

6.6. Maintenance

Cells were grown in a carbon dioxide (CO_2) cell culture incubator (HERAcell cell culture incubator, Heraeus, Hanau, Germany) at 37 °C, 5% CO_2, and with more than 95% relative humidity.

6.6.1. Freezing

Cells were collected from cell culture plastic, washed once in PBS, resuspended in 90% fetal calf serum 10% dimethyl sulfoxide (DMSO), and aliquotted to CryoTubes™ (Nunc, Thermo Fisher Scientific). CryoTubes were placed in a styrofoam box at -70 °C to allow for optimal freezing at -1 °C/min. After one to three days, cells were transferred to liquid nitrogen (N_2).

6.6.2. Restimulation

T cell lines and clones were regularly restimulated with feeders and phytohemagglutinin (Remel, Thermo Fischer Scientific, Lenexa, Kansas, USA). Restimulation was done in less than 4 d old HSM supplemented with 1×10^6 per ml 3000 cGy (rad) irradiated PBMC as feeders and 2 µg/ml phytohemagglutinin. 24-well plates with T cells ranging from 2×10^5 to 2×10^6 in 1 ml were mixed with 1 ml of restimulation medium. Intervals of at least 2 weeks before proceeding to a subsequent restimulation were kept.

6.6.3. Thawing

The freezing tube was taken from liquid nitrogen via dry-ice and placed into a 37 °C water bath until only a small piece of ice was left. Cells were pipetted to a 15 ml conical tube containing 2 ml PBS and once 6 ml PBS were added dropwise while gently shaking the tube. Thawed cells were centrifuged 7 min at 1000 rpm (209 g) and 15 °C. The supernatant was discarded and the tube was scratched to loosen the pellet. Once 8 ml PBS were added slowly shaking the tube concurrently. Cells were spun 7 min at 1000 rpm (209 g) and 15 °C in a bench top centrifuge. The supernatant was discarded and the tube scratched to loosen pellet. Cells were plated into an appropriate number of 6-well or 24-well plates.

6.7. Sorting

Cells and controls were stained in PBS containing 0.1% human albumin and 2 mM EDTA if staining was not affected. DNA intercalating agents such as propidium iodide (PI) or 7-aminoactinomycin D (7AAD) were avoided because of their detrimental effects on cells. 20 µg/ml gentamycin (Gibco) was used additively to prevent contaminations. All stained cells and controls plus single stains for compensation were collected through a cell strainer (sieve of nylon mesh to retain cell clumps) to FCSM-coated polypropylene tubes for sorting. The volume was not to deceed 800 µl and cell densities had to range from 1 to 3×10^7 cells per ml. Sorted cells were collected in FCSM-coated tubes containing a bottom layer of FCSM plus gentamycin.

7. Immunological techniques

7.1. Antigen presentation assay

All types of antigen presentation assays described below were performed as cell-free stimulation assays by replacing the presenting cells with plate-bound antigen-presenting molecules either as monomers, dimers or multimers.

7.1.1. Antigen presentation assay, chase

Presenting cells (2.5×10^4 per well) were pulsed for 1 h with sonicated antigen at 37 °C in 10% FCSM (see section 5.6.1, p.39), washed, and chased for different time periods before addition of T cells at the same time point (1×10^5 per well). The supernatants were harvested after 24 or 48 h and released cytokines were measured by ELISA (see section 7.7, p.46).

7.1.2. Antigen presentation assay, competition

Presenting cells (2.5×10^4 per well) in 10% FCSM (see section 5.6.1, p.39) were incubated during the whole assay at 37 °C with sonicated antigen (at the indicated concentrations). Competing compounds were given several hours in advance of αGC. T cells (7.5×10^4 per well) were added 30 min after addition of αGC. Supernatants were harvested after 24 or 48 h and cytokine release was assessed by ELISA.

7.1.3. Antigen presentation assay, pulse

Presenting cells (2.5×10^4 per well) in 10% FCSM (see section 5.6.1, p.39) were pulsed for the indicated time periods with sonicated antigen at 37 °C. Presenting cells were washed extensively and T cells were added (1×10^5 per well). The supernatants were harvested after 24 or 48 h and released cytokines were measured by ELISA.

7.1.4. Antigen presentation assay, standard

Presenting cells (2.5×10^4 per well) in 10% FCSM (see section 5.6.1, p.39) were pulsed for 1 h and incubated during the whole assay, at 37 °C, with sonicated antigen. After 60 min, T cells were added (1×10^5 per well). The supernatants were harvested after 24 or 48 h and released cytokines were measured by ELISA.

7.2. Cell tracing and tracking

Washed cells and dyes were prepared in warm medium. Cells were left for 5 min to take up 0.1 µM carboxyfluorescein succinimidyl ester (CFSE) in a cell culture incubator (see section 6.6, p.43) at 37 °C or with 1 µg/ml 4',6-diamidino-2-phenylindole dihydrochloride (DAPI) (Sigma-Aldrich) or Hoechst (e.g. 33342, Sigma-Aldrich) for 30 min. The labeling was stopped by addition of warm PBS containing 5% human albumin to absorb the dye. Labeled cells were washed in warm PBS before usage.

7.3. Co-culture assay

Labeled cells of interest were placed on top of a feeder cell layer or APC and given five to seven days until analysis of co-culture effects by appropriate means.

7.4. Cytotoxicity assay

Compounds to be tested were sonicated for 5 min in a Elmasonic X-tra 150 H sonicator (Elma, Singen, Germany) at 150 Hz 37 °C and incubated overnight at increasing doses with cells. The next day, cells were labeled with 5 µg/ml PI (Sigma-Aldrich) or 7AAD (Invitrogen). Cell death, measuring PI or 7AAD uptake into the cells, was assessed by fluorescence-activated cell sorting (FACS) (see section 7.8, p.47) on a CYAN^{TM} ADP (Beckman Coulter, Nyon, Switzerland). The median lethal concentration (LC_{50}) were calculated for tested compounds.

7.5. Immunoprecipitation

Immunoprecipitation of proteins was performed on protein G sepharose beads (GE Healthcare, Otelfingen, Switzerland). Washed cells (more than 1×10^6 per immunoprecipitation) were lyzed by repeated freeze-thaw cycles or a suiting detergent (Triton-X100, Brij98, etc.) containing lysis buffer with proteinase inhibitor cocktail and 1 mM phenylmethylsulfonyl fluoride for 1 h on ice. 1 volume of 100 mM Tris pH 7.2 was added to lyzed pellets. One tenth volume of washed protein G sepharose beads was added for preclearing the lysate and rotated 1 h overhead on a wheel at RT. Lysate supernatant was transferred and incubated with 10 µg antibody in a maximal total volume of 500 µl overnight on an overhead rotating wheel at 4 °C. One tenth volume of washed protein G sepharose beads was added for immunoprecipitating the lysate and placed for 3 h on an overhead rotating wheel at 4 °C. Beads were precipitated by centrifugation and the immunoprecipitation was washed twice before Western blotting.

7.6. Confocal laser scanning microscopy (CLSM)

Fresh-frozen, OCT-embedded tissue sections or cells, either grown or fixed and cytospun or cytospun on polylysine-coated slides, were fixed in 2% PFA at RT or 100% acetone -20 °C. In case of future use non- or fixed preparations were frozen at -70 °C. After fixation cells were permeabilized in saponin buffer (see section 5.6.2, p.39). Blocking was done with CAS block (Invitrogen) and possibly appropriate non-immune immunoglobulins. Samples were stained with antibodies against the marker of interest followed by a final nuclear counterstain (e.g. DAPI, Hoechst or PI). Slides were mounted in fluorescence mounting medium (Dako, Beckman Coulter), sealed with nail polish and stored at 4 °C until analysis. Analyses were performed on a Zeiss LSM510 or LSM710 confocal laser scanning microscope (Carl Zeiss, Feldbach, Switzerland).

7.7. Enzyme-linked immunosorbent assay (ELISA)

MaxiSorp^{TM} ELISA plates (Nunc, Thermo Fisher Scientific) were coated with capture mAb diluted in PBS overnight at 4 °C. The plate was blocked with ELISA blocking solution and incubated with the supernatants of the antigen presentation assays. Detection was performed with the biotinylated partner mAb in ELISA blocking solution. Streptavidin-horseradish peroxidase (Zymed, Invitrogen, Basel, Switzerland) with o-Phenylenediamine dihydrochloride (Sigma-Aldrich, according to the manufacturer's instructions) as substrate were used to develop the

ELISA. The reaction was stopped by adding half a volume of 10% (v/v) sulphuric acid (Sigma-Aldrich). Opitcal densities of the developed wells were read in a SpectraMax® 190 spectrophotometer at 490 nm (Molecular Devices, Sunnyvale, California, USA) and converted to concentrations expressed as mean pg/ml ± standard deviation (SD) of triplicates or quadruplicates using the SoftMax Pro 5 program (Molecular Devices) by comparison to recombinant standards.

7.8. Fluorescence-activated cell sorting (FACS)

Fixed or non-fixed cells were plated to a 96-well U-bottom plate. Fixed cells were processed in saponin buffer (see section 5.6.2, p.39) for intracytoplasmic staining at RT whereas living cells were handled on ice with FACS buffer (see section 5.6.2, p.39).

Fluorescent labels were chosen according to the expression of the markers of interest. Weak markers were given bright labels and strong markers were given duller labels. In multicolor analysis channels of the flow cytometer were chosen to minimize spectral overlap of the fluorochromes used.

Antibodies (see section 5.1, p.31) were used in general in the range of 0.2 to 20 µg/ml. Titrations of all used antibodies were performed on appropriate target cells to achieve optimal staining to background (signal to noise) ratios.

Cells were acquired on a CYAN ADP flow cytometer (Dako, Beckman Coulter), and events were gated to exclude non-viable cells on the basis of light scatter and PI or 7AAD incorporation as well as on pulse-width of the forward scatter signal to include only single living cells.

7.9. Fluorescence microscopy

Cells were treated as for confocal laser scanning microscopy (see section 7.6, p.46) but analyzed on a Olympus BX61 fluorescence microscope (Olympus, Volketswil, Switzerland).

7.10. Proliferation assay

7.10.1. Thymidine uptake

Was replaced by CFSE (see subsection 7.10.2, p.47) whenever possible.

7.10.2. Carboxyfluorescein succinimidyl ester (CFSE)

Cells were marked with CFSE (see section 7.2, p.45) and activated or subjected to a co-culture assay (see section 7.3, p.46) with stimulatory cells. Antigen was added to assess antigen-specific proliferation of a population of interest.

8. Molecular techniques

8.1. Genomic deoxyribonucleic acid (gDNA) preparation

DNA was extracted from 1 to 2×10^6 cells, left in culture for more than 1 week after thawing, with NucleoSpin Blood (Macherey-Nagel, Oensingen, Switzerland) as per manufacturer's protocol (see appendix SOP mycoplasma, p.199). Eluted DNA was stored correspondent to NucleoSpin Blood recommendations (e.g. -20 °C).

8.2. Electrophoresis

8.2.1. Agarose gel electrophoresis

For all screening or test PCR products a 1% agarose gel was run and digital pictures were taken while exposing the intercalating ethidium bromide to ultraviolet light.

8.3. Polymerase chain reaction (PCR)

For all screenings or tests primers were selected and a standard cycling program was established to be run with Taq polymerase (see appendix SOP mycoplasma, p.199).

8.4. Vesicle uptake studies

Blood was drawn from healthy donors into heparinized syringes and immediately processed. One-step inside-out membrane vesicles (OSV) were prepared as previously described [Wu et al. 2005]. ATP-dependent transport of radioactive substrates into OSV was measured by a rapid filtration technique. For ^{3}H guanosine 3',5'-cyclic monophosphate (cGMP) control and ^{3}H IPP or ^{14}C IPP (all ARC, St. Louis, Missouri, USA) uptake studies, OSV were used in a buffered system (55 µl total volume) containing 1 mM ATP, 10 mM $MgCl_2$, 10 mM creatine phosphate, 100 µg/ml creatine phosphokinase (all Sigma-Aldrich), 10 mM Tris−HCl (pH 7.4) and 3.3 µM radioactive substrate. After 60 min at 37 °C incubation, 1 ml of ice-cold 10 mM Tris−HCl (pH 7.4) was added and thereof 0.5 ml was filtered through 3% (w/v) bovine serum albumin soaked polyvinylidene fluoride filters (Millipore, 0.22 µm pore size). The filters were rinsed three times with 2 ml 10 mM Tris-HCl (pH 7.4) and the radio-labeled substrate retained on the filter was determined by liquid scintillation counting in Pico-Fluor™ 15 scintillation liquid (Perkin Elmer, Waltham, Massachusetts, USA). Obtained radioactivity values were expressed as counts per minute (cpm). ATP-dependent uptake is the difference in tracer retention on the filter with ATP-γ-S (a non-hydrolyzable ATP analogue, Sigma-Aldrich) in the presence of the ATP regenerating system (10 mM creatine phosphate, 100 µg/ml creatine phosphokinase), compared to values obtained with ATP.

9. Patients and healthy donors

All investigations with human subjects and tissues were approved by the regional ethical review boards and were performed in accordance with institutional guidelines.

9.1. Atherosclerosis patient cohort from Cantonal Hospital Luzern

Cells and plasma of a random cohort of 350 persons seeing a doctor for manifold reasons were collected in compliance with the SOP for isolation of PBMC (see appendix SOP PBMC, p.195) developed for our collaborators at the cardiology unit of the Cantonal Hospital Luzern.

Table 9.1 Clinical characteristics of the Lucerne patients

Group [i]	1	2	3	6
Individuals	(67)	(96)	(192)	(81)
Continuous variables [ii]				
Age [iii]	69 ± 10 (67)	58 ± 11 (96)	62 ± 9 (192)	54 ± 12 (81)
BMI [iv]	26.1 ± 4.4 (66)	27.0 ± 4.1 (95)	26.6 ± 4.1 (188)	26.9 ± 3.9 (80)
Leuk [v]	6.9 ± 2.6 (67)	7.7 ± 3.6 (94)	6.7 ± 2.6 (187)	6.0 ± 2.3 (79)
LDL [vi]	1.9 ± 0.8 (48)	2.8 ± 1.0 (83)	2.4 ± 1.1 (185)	2.7 ± 1.0 (76)
HDL [vi]	1.1 ± 0.4 (48)	1.2 ± 0.4 (87)	1.3 ± 0.6 (188)	1.4 ± 0.5 (79)
Tgl [vi]	1.3 ± 1.0 (48)	1.3 ± 1.1 (87)	1.3 ± 1.0 (187)	1.2 ± 1.3 (79)
Chol [vi]	3.9 ± 1.0 (50)	4.9 ± 1.3 (88)	4.5 ± 1.2 (188)	4.8 ± 1.1 (79)

continued on next page

[i] 1: Scheduled for vascular surgery (histologically proven PAD/CerebroVD), 2: Acute CAD (ACS included unstable AP), 3: Chronic CAD (relevant coronary stenosis), 6: No significant coronary atherosclerosis (but endothelial dysfunction, small vessel disease, coronary sclerosis)
[ii] All continuous variables are given as median ± SD (individuals measured)
[iii] Given in years
[iv] Body mass index given in kg/m^2
[v] Leukocyte count given in 1×10^9/l
[vi] Low-density lipoprotein (LDL), high-density lipoprotein (HDL), triglyceride (Tgl) and total cholesterol (Chol) given in mmol/l

continued from previous page

Group [i]	1	2	3	6
Individuals	(67)	(96)	(192)	(81)
Categorical variables [vii]				
Male gender	53 (67)	75 (96)	143 (192)	42 (81)
Diabetes	11 (67)	12 (96)	37 (192)	7 (81)
Hypercholesterolemia	40 (67)	59 (96)	142 (192)	36 (81)
Hypertension	54 (67)	46 (96)	124 (192)	33 (81)
Smoker	52 (67)	57 (96)	107 (192)	30 (81)
Family history	19 (67)	33 (96)	67 (192)	32 (81)
CAD [viii]	21 (29)	92 (96)	185 (191)	7 (81)
PAD [ix]	28 (67)	1 (96)	12 (192)	3 (81)
CVI [x]	21 (67)	4 (96)	7 (192)	1 (81)
COPD [xi]	12 (67)	3 (96)	8 (192)	3 (81)
Aneurysm	23 (67)	1 (96)	3 (192)	0 (81)
ACBP [xii]	8 (67)	2 (96)	3 (192)	0 (81)
Collagenosis	2 (67)	1 (96)	3 (192)	2 (81)
PAT [xiii] elevated	21 (61)	14 (85)	51 (177)	35 (79)
TCFA [xiv]	1 (1)	20 (23)	64 (92)	5 (36)

[i] 1: Scheduled for vascular surgery (histologically proven PAD/CerebroVD), 2: Acute CAD (ACS included unstable AP), 3: Chronic CAD (relevant coronary stenosis), 6: No significant coronary atherosclerosis (but endothelial dysfunction, small vessel disease, coronary sclerosis)
[vii] All categorical variables are given as number (individuals measured)
[viii] Coronary artery disease (CAD) or atherosclerotic heart disease
[ix] Peripheral vascular disease (PVD), also known as peripheral artery disease (PAD) or peripheral artery occlusive disease (PAOD)
[x] Cerebrovascular insult (CVI) or stroke
[xi] Chronic obstructive pulmonary disease (COPD)
[xii] Aorto-coronary bypass graft (ACBP)
[xiii] Peripheral artery tonometry (PAT), non-invasive endothelial function measurement (score<1.7 abnormal, score>2.0 normal (means good endothelial function))
[xiv] Thin-capped-fibro-atheroma (TCFA) refers to a special form of plaque composition which is prone to rupture

Additionally, clinical parameters were assessed by our collaborators whenever possible and after having obtained written consent from the examined subject. Data collected were basic information (e.g. body measures, sex, race, etc.), reasons of hospitalization, risk factors (e.g. diabetes, hypercholesterolemia, hypertension, smoking, family history, etc.), manifestations of ATH, comorbidities, actual/former drug therapy, laboratory blood test results, angiographic parameters, flow measurements and intravascular ultrasound (IVUS) of the left/right coronary artery, of the left anterior descending artery and of the circumflex artery.

9.2. Atherosclerosis patients from Cantonal Hospital Bruderholz

Table 9.2 Clinical characteristics of the Bruderholz patients

Cardiovascular events Individuals	No (n = 27)	Yes (n = 22)	P value
Cardiovascular risk factors			
Diabetes mellitus [i]	1 (4)	10 (45)	<0.001
Hypercholesterolemia [i]	2 (7)	7 (32)	0.03
Arterial hypertension [i]	6 (22)	11 (50)	0.04
Smoking [i]	5 (18)	7 (32)	0.35
Male sex [i]	13 (48)	11 (50)	0.90
Age [ii]	74 ± 14	79 ± 9	0.12
Body mass index [iii]	23 ± 6	28 ± 6	0.01
History of cardiovascular disease			
Coronary heart disease [i,iv]	0 (0)	21 (95)	<0.001
Cerebrovascular disease [i,v]	0 (0)	8 (36)	<0.001
Arterial occlusive disease [i,vi]	0 (0)	8 (36)	<0.001
Infection at death [i,vii]	11 (41)	10 (45)	0.80
Autopsy [viii]	20 ± 12	24 ± 9	0.28

[i]   Number (%)
[ii] Mean years ± SD
[iii] Mean kg/m^2 ± SD
[iv] Myocardial infarction, angina pectoris with myocardial ischemia, revascularization
[v]   Cerebrovascular ischemic stroke, transient ischemic attack, revascularization
[vi] Symptomatic peripheral arterial occlusive disease, symptomatic aortic aneurysm, revascularization
[vii] Infection at death was defined by the presence of two or more of the following criteria: body temperature >38 °C, C-reactive protein >5 × 10^4 mg/ml, neutrophils (band forms) >10%, positive blood cultures
[viii] Mean hours after death ± SD

Arterial ring segments were obtained during autopsy from deceased patients who were treated for a broad variety of medical conditions at the department of general medicine of the academic medical center Cantonal Hospital Bruderholz [Kyriakakis et al. 2009].

9.3. Healthy donor cohort from Blutspende beider Basel

Buffy coats were received fresh after preparation by the unit in charge of the University Hospital Basel, Blutspende beider Basel, and cells were immediately prepared as described above (see subsection 6.4.2, p.42). The overall time required from the donation of blood to a purified PBMC population was no more than 6 h.

10. Software and statistical analyses

10.1. CLSM software

10.1.1. ImageJ

ImageJ was developed for Image processing and analysis in Java http://rsbweb.nih.gov/ij/.

Images were loaded into ImageJ [Rasband 1997–2009] and visualized using HiLo LUT for background correction (thresholding). Images were quantified by the Intensity Correlation Analysis (ICA) [Li et al. 2004] and Colocalisation Thresholds (CT) [Costes et al. 2004] plugins. At least four complete images containing more than 8 single cells without region of interest (ROI) selection were used to calculate Mander's Colocalization Coefficients for channel 1 (M1: e.g. red overlapping with green) and channel 2 (M2: e.g. green overlapping with red) and the Intensity Correlation Quotient (ICQ: random staining ICQ $\approx$ 0, segregated staining $0 > $ ICQ ≥ -0.5, dependent staining $0 < $ ICQ $\leq +0.5$). Colocalized pixels were calculated using the Colocalization Highlighter (CH) plugin with the median channel thresholds of channel 1 and channel 2 of all images of a specific marker established by the CT plugin and the results are shown as binary image.

For channels containing autofluorescence, a median filter with 1 pixel size was applied if stated. For 3D-reconstruction overlapping Z-stacks were acquired and processed with 5 or 10 degree angle increment with interpolation. Scale bars of 10 µm were placed in images if not denoted otherwise. Final assembly of the images was done with the ImageJ internal montage plugin.

For every project adequate macros were developed and applied to the images of choice to guarantee unbiased and equal processing. Standardized batch runs were performed on selected sets of dimension-matched images.

10.1.2. Imaris

http://www.bitplane.com/

10.1.3. ZEN

http://www.zeiss.com/c12567be0045acf1/Contents-Frame/3f3821b370efc91cc125734c002fb38c/

10.2. FACS software

10.2.1. FCS Express

http://www.denovosoftware.com/site/FCSExpress.shtml/

FCS Express was used to create publication grade flow cytometry plots based on the preanalysis done in Summit and exported as jpg-file.

10.2.2. FlowJo

http://www.flowjo.com/

For analyses needing batch processing, fcs-files with stored compensation performed in Summit were subjected to further evaluation with FlowJo.

10.2.3. Summit

The Summit software was developed by Dako and used to acquire FCS-files on a CYAN ADP flow cytometer.

FCS-files were analyzed using the Summit software and if applicable compensation was performed with single stains of used fluorochromes. Cells of interest were chosen according to their forward- and side-scatter profile and further gated on pulsewidth versus a dead cell exclusion dye to only include single living cells in the analysis.

10.3. Fluorescence microscopy software

10.3.1. cell^P

The Olympus cell^P software (http://www.olympus.de/microscopy/35_cell_P.htm/) was used for image acquisition on an Olympus BX61 fluorescence microscope.

10.4. Graphic software

10.4.1. GIMP

GIMP is the GNU Image Manipulation Program. Further information can be found at http://www.gimp.org/.

10.5. Line art software

10.5.1. GraphPad Prism

GraphPad Prism was used for basic biostatistics, curve fitting and scientific graphing, especially line art. For further information on GraphPad Prism visit http://www.graphpad.com/prism/Prism.htm/.

10.6. Statistical software

10.6.1. GraphPad Prism

See the subsection on GraphPad Prism in the line art section above for information on the program.

10.6.2. R

R is a free software environment for statistical computing and graphics and was used for more complex statistical data evaluation not available within the GraphPad Prism Software. For further information on the R package visit http://www.r-project.org/.

Part III.

Results and Discussion

11. CD1-DIMER, a T cell envision tool

Historically, it has been a challenge to directly identify T cells for their specificities. T cell receptor (TCR) affinity is intrinsic to its amino acid sequence and three-dimensional structure but generally is low. A solution to stain for the TCR was first reported using multimerization of peptide-major histocompatibility complex (MHC) molecules thereby augmenting the avidity by multiple simultaneous interactions. Later, CD1d tetramers were employed to detect invariant natural killer T (iNKT) cells. Firstly, every presenting molecule needs to be identically loaded; ensured when handling a synthetic antigen, but ligand preparations from tissue are hardly pure enough to fulfill the criteria. Secondly, the usually low fluorescence intensity hinders an unambiguous identification of positive cells. The CD4 and CD8 co-receptors, providing a substantial part of the binding energy for peptide-MHC, do not interact with CD1d and therefore can be neglected when staining iNKT cells.

To circumvent these drawbacks, CD1d was engineered to be dimerized and potentially multimerized on polymers at will. Visualizing iNKT cells in peripheral blood by alpha-galactosylceramide (αGC)-CD1d dimers instead of tetramers unveiled huge differences in fluorescence intensity. Two populations, defined as dull and bright, were sorted and cloned to assess possible functional dependencies. More than eighty iNKT cell clones were phenotypically and functionally examined. Comparative analyses did neither show correlation to half maximal activation by αGC nor any activatory, inhibitory and costimulatory molecules. Considering that a co-receptor for iNKT cells has not been identified so far and the invariance of the TCR alpha chain, the TCR beta chain should dominate the T cell specificity. More recent structural data confirmed the importance of the TCR beta chain sitting on top of the F' pocket of CD1d that can induce conformational changes in CD1d when loaded with different types of antigens.

Peptide-MHC tetramer-induced activation requires T cell co- and receptor binding but not co-stimulation; αGC-CD1d dimer-induced activation needs TCR triggering only. As to this singular prerequisite, activation with plastic-bound antigen-loaded CD1d dimers was performed. Chemically designed and synthesized piperidinones unexpectedly revealed that a single lipid tail is sufficient to form stimulatory complexes with CD1d and to induce cytokine release by iNKT cells. All active piperidinones preferentially induce T_H1-like cytokines, whereas the T_H2 prototype cytokine IL-4 is released to a lesser extent. As both classes of cytokines are detected after αGC stimulation, the T_H1-bias has to be imputed to the type of lipid-CD1d complexes formed by the piperidinones. T_H1-responses have been associated with prolonged TCR engagement, hence piperidinones might form complexes with CD1d that are binding with high affinity to the TCR of iNKT cells. Structural fine-tuning of piperidinones may lead to new, unique biological properties deployable in vaccination and anti-tumor therapies.

Comprehension of the biological roles of T cells had been hampered by the lack of tools to determine the frequency and phenotype of antigen-specific T cells. Up to recently, antigen-specific T lymphocytes had been identified mainly by functional assays as cytokine secretion, cytotoxic activity or proliferation — usually performed after culture of the cells *in vitro*. Over the past years, the development of tetrameric peptide-loaded major histocompatibility complex (MHC) class-I complexes has radically altered the way of monitoring T cells and their responses.

In order for a T cell to react to the peptide or lipid antigen to which it is specific, it has to sense the antigen in the context of the antigen-presenting molecule (APM) complex, MHC or CD1, respectively. Even though being specific, T cell receptors (TCR) are known to have low affinities for their cognate antigen-APM counterparts, with combined fast dissociation rates and short half-lives in the range of a few seconds. The interactions between TCR and monomeric antigen-APM are therefore too labile to be exploited to effectively label T cells. But joining multiple APM molecules into one complex greatly enhances binding stability. In a hallmark paper [Altman et al. 1996], a first solution, how to stain T cells for their specificities, was introduced for MHC-restricted peptide-specific T cells. To approach the need for multivalency, multimers of MHC molecules were designed to overcome the low affinity of a single TCR by increasing the overall avidity of the interaction by multiple concomitant bindings. Therefore biotinylated soluble MHC monomers were bound to (fluorochrome-conjugated) streptavidin, having four biotin binding sites, and thereby converted to tetravalent structures. The resulting MHC tetramers remain the most widely used reagents to detect antigen-specific $CD4^+$ and $CD8^+$ T cells by flow cytometry [Altman & Davis 2003; Bakker & Schumacher 2005; James et al. 2009; Wooldridge et al. 2009].

Typically, MHC class I tetramers are formed by refolding soluble recombinant MHC class I heavy chain and β_2 microglobulin (β_2m) obtained from bacterial inclusion bodies in the presence of a single peptide. While this approach is successful in yielding homogeneously loaded MHC class I molecules, attempts to refold CD1 molecules by this method have failed. That failure may be due to the larger amount of hydrophobic amino acids in CD1 than MHC class I molecules, needed to bind glycolipid antigens. Lately, two protocols were described for the generation of CD1d tetramers. One is based on CD1d molecules derived from insect cells [Benlagha et al. 2000; Matsuda et al. 2000], while the other is based on denatured CD1d molecules that are refolded *in vitro* by oxidative chromatography [Karadimitris et al. 2001].

CD1d tetramer staining assays were initially reported for mouse invariant natural killer T (iNKT) cells [Gumperz et al. 2000; Matsuda et al. 2000] with some cross-reactivity to *in vitro* alpha-galactosylceramide (αGC)-expanded human iNKT cells [Benlagha et al. 2000]. A similar strategy successfully identified human iNKT cells by a tetramerized αGC-loaded engineered human β_2m-CD1d-Fc fusion protein [Gumperz et al. 2002] in fresh peripheral blood mononuclear cells (PBMC). Akin technologies are currently used to detect the presence of iNKT cells in mouse and man.

Contrasting reports found that several lipid antigens could tightly bind (i.e. with a slow dissociation rate) to mouse CD1d as read out by T cell hybridomas [Benlagha et al. 2000] or found a very rapid off rate with a half-life of αGC-CD1d complexes in the order of 7 s to 3 min by plasmon resonance [Naidenko et al. 1999]. That contradiction may be related to technical issues associated with BIAcore technology and/or the use of a biotinylated form of αGC with potentially changed CD1d1 binding properties.

While only CD1d fusion proteins were used in tetramer-based staining assays, we devised a system based on CD1d molecules, naturally (non-covalently) bound to β_2m, that are tagged to be dimerized by an anti-tag mAb and that can be loaded by the antigen of interest (see section 5.5, p.37). The final construct (see Figure 11.1, p.62) was designed to avoid any conformational or structural constraints that could be introduced by fusion protein approaches and, therefore was designed to allow for conformational changes in the CD1d protein upon antigen binding, exposure

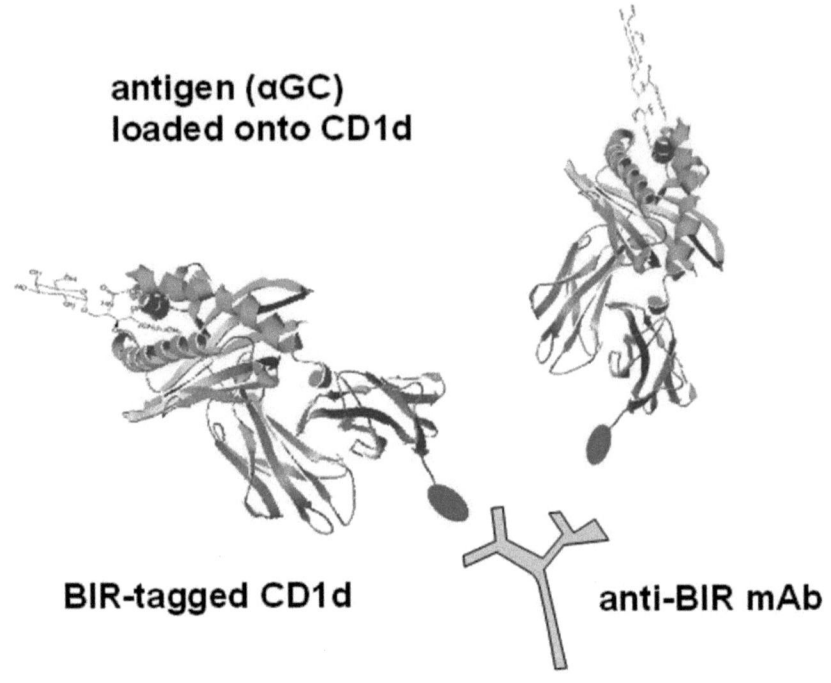

Figure 11.1
Alpha-galactosylceramide-loaded human CD1d DIMER. Soluble recombinant human CD1d is loaded with the archetype iNKT-ligand αGC and dimerized with the BIR1.4 mAb specific for the birA-tag engineered at the C-terminal end of the CD1d protein.

to different environmental pH or TCR interaction. Our β_2m-associated soluble recombinant human CD1d loaded with αGC and dimerized with the tag-specific BIR1.4 mAb is referred to as DIMER.

As loaded monomers of CD1d are not able to stain iNKT cell clones (data not shown), dimerization was carefully controlled by diluting the BIR1.4 mAb against CD1d. The testing demonstrated the expected sensibility of the system to the ratio of CD1d to BIR1.4 (see Figure 11.2 (A), p.63). In order to establish the sensitivity of the novel system an antigen titration was performed using the iNKT cell antigen αGC, starting from a molar ratio of one to roughly two hundred of CD1d to αGC (see Figure 11.2 (B), p.63). The required amounts of antigen for reliable and optimal staining by the DIMER was assessed on the iNKT cell clone JS7. By diluting the first plateau concentration of αGC six times by twofold the JS7 population shifted to background MFI levels as seen by FACS analysis (see Figure 11.2 (B), p.63). Molar excess of the candidate antigen is of critical importance especially for low affinity ligands as the available amount of antigen for loading remains deceptive due to lipid behavior in aqueous solution (*inter alia* micellization).

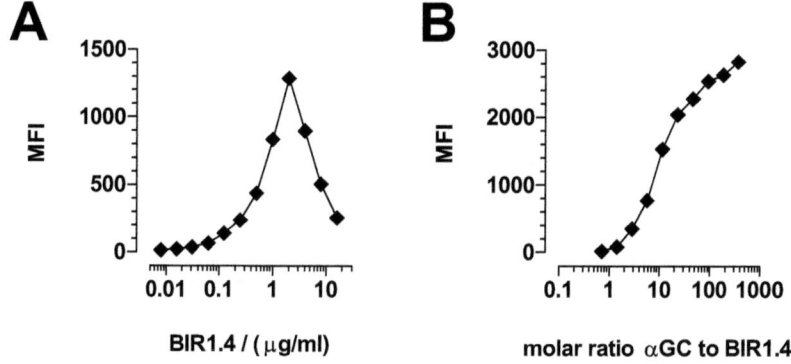

Figure 11.2
Establishing the DIMER system. (A) Titration of BIR1.4 mAb against αGC-loaded monomers and subsequent staining of the iNKT cell clone JS7 showed that the sensibility changes in a fourfold span around the optimum with a reduction in MFI to half by increasing or decreasing the BIR1.4 concentration by twofold. (B) In the region of a five to fifty molar ratio of αGC to CD1d (10-100 αGC to BIR1.4) the DIMER staining rises from substantially different from background to plateau.

Every batch of DIMER was carefully tested and analyzed before usage to assure identical conditions of this crucial tool. Higher order multimers (trimers, tetramers, pentamers, 'dextramers' and polymers) will have longer interaction half-lives at the cell surface under staining conditions around the zero point. However, when staining at physiological temperatures or RT, multimers are swiftly internalized [Whelan et al. 1999]. Under the latter conditions, a potential advantage of prolonged multimer dwell times is outbalanced by cellular uptake. Utilizing the DIMER multimerized on dextran could abate the requirement of highly pure antigen, therefore tests were conducted on selected iNKT cell clones showing different levels of DIMER staining. The effect of dextran-DIMER multimers is valid over a broad range of staining intensities and furthermore under limiting amounts of antigen the MFI can be augmented as much as 80-fold compared to the DIMER (data not shown). Therefore and in addition because the CD1d molecules may already be loaded by lipids during production and purification, it is likely that the dextran-DIMER multimers could be used for staining even mixtures of antigens or tissue sample lipid preparations.

11.1. Clonal brightness

Comparing the DIMER staining of a cohort of iNKT cell clones to their surface expression of total TCR (by means of anti-CD3 mAb) or of the TCR variable chain regions (by anti-TCR Vα24 and anti-TCR Vβ11 mAb), we repeatedly noticed a discrepancy between TCR or TCR variable chain expression and DIMER staining (see Table 11.1, p.64).

11. CD1-DIMER, a T cell envision tool

Table 11.1 Clonal brightness (MFI)

iNKT cell clone	CD3	TCR Vα24	TCR Vβ11	DIMER
JS7	90	583	673	172
JS11	63	379	487	104
JS63	78	392	698	284
LMP-D9	60	340	422	104
VM-D5	50	340	422	453

Moreover, the clone with the lowest surface expression of the TCR showed the most intense labeling by DIMER. All murine iNKT cell hybridomas contain a characteristic single N region nucleotide addition at the V-J junction of their TCR Vα14 chain to preserve the glycine and aspartate at residues 93 and 94, respectively [Behar et al. 1999]. The human TCR Vα24 to Jα18 junction contains the entire Jα18 gene segment except for two cytosines immediately after the heptamer recombination signal and thus is created by deletion of two bases without N region additions [Porcelli et al. 1993]. Therefore, a varying contribution of the TCR α-chain can be excluded and the diverse (DIMER) staining could possibly be explained by a co-receptor being differentially expressed on the clones or by a vast contribution of the TCR β-chain as well as its P- and N-nucleotides to the total binding footprint. Mouse iNKT cells preferentially use several TCR Vβ genes (e.g. Vβ8.2 or less frequently Vβ7 or rarely Vβ2) for their TCR Vα14 chain, whereas human iNKT cells pair their TCR Vα24 mostly with the Vβ11 chain; other TCR Vβ chains are hardly ever found on mouse and human iNKT cells.

This led us to derive αGC-expanded lines from PBMC of several healthy donors, to sort them for dull or bright DIMER staining (see Figure 11.3, p.65) and to clone the sorted populations. The clones are named according to their origin **B**uffy coat-derived α**GC**-expanded from a plating with 0.3 cell per well (**A**) or with one cell per well (**B**). In addition, the more than eighty clones are numbered for their DIMER MFI (f.i. BGA1 or BGB24) with 1 being the dullest and 89 being the brightest clone (see Figure 11.5, p.66). The dull sorted iNKT cells seemed to be minimally more efficient than bright sorted ones (see Figure 11.4, p.66). Reanalysis by DIMER staining of the BG clones confirmed the reliability of the DIMER technique as most of the dull sorted clones stayed duller than the bright sorted clones (see Figure 11.5, p.66).

The BG clones did not show any restraints in expression of CD4 and CD8 as their phenotypes included double-negative (DN), CD4$^+$ and CD8$^+$ cells. Only one clone (CD8$^+$) was TCR Vα24- and Vβ11-negative but it could not be stained by the DIMER [Behar & Cardell 2000].

Whereas TCR Vα24 significantly (P=0.0175) correlated with TCR Vβ11 (data not shown), as expected, neither TCR-chain could be directly linked to DIMER staining. When comparing the quotients DIMER divided by TCR Vα24 and DIMER divided by TCR Vβ11 the correlation was highly significant (P<0.0001) (see Figure 11.6, p.67). This fact seems to point to a sole contribution of the TCR chains to the antigen recognition, nevertheless a contribution of a to date unknown co-receptor can not virtually be excluded. No direct correlation could be seen between the classical TCR co-receptor molecules CD4 or CD8 and DIMER staining. A skewed picture became obvious when the DIMER staining was subdivided into dull, intermediate and bright. Dull clones tend to express CD8, on the other hand bright clones contained more CD4-positive ones than dull (and intermediate) clones. DN clones were found in all DIMER intensity populations (see Figure 11.7, p.67).

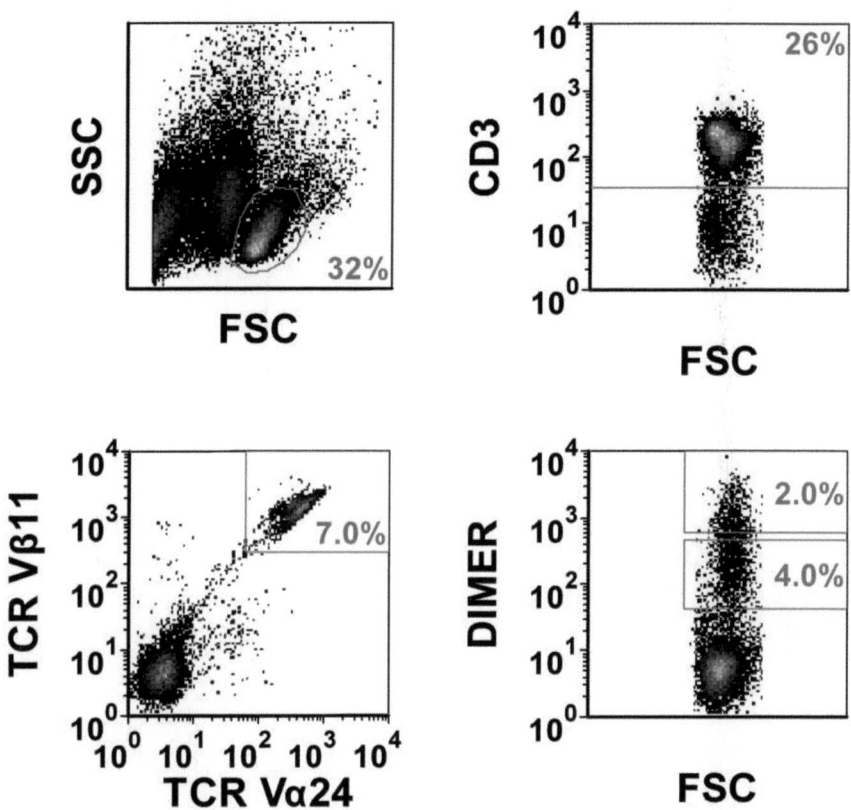

Figure 11.3
Sorting of dull and bright DIMER-positive iNKT cells from PBMC. PBMC were stained for CD3, TCR Vα24, TCR Vβ11 and DIMER. Cells were consecutively gated on pulsewidth and propidium iodide (PI) (not shown), forward (FSC) and side scatter (SSC), and CD3. Then TCR Vα24 was plotted against TCR Vβ11 and FSC against DIMER. Both ways to visualize iNKT cells showed 6-7% of total T cells presented an iNKT cell phenotype after αGC expansion. Percentages are given of total acquired cells.

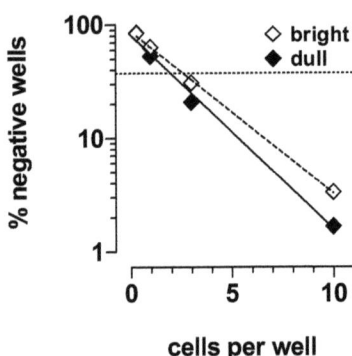

Figure 11.4
Plating efficiency of the BG clones. Overall plating efficiency was very high with two to three cells per well giving a clone. The dull sorted iNKT cells (filled diamonds) were slightly more effective than the bright sorted ones (empty diamonds). The dotted line shows the 37% cut-off.

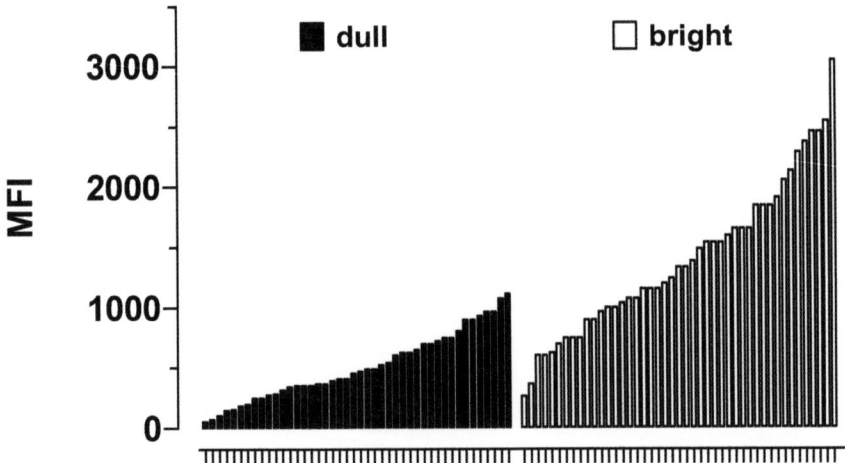

Figure 11.5
Clonal brightness. Reanalysis by DIMER staining of the BG clones confirmed the reliability of the technique and led to the naming of the BG clones according to DIMER MFI ranking with 1 given to the dullest and 89 to the brightest clone. MFI of dull sorted clones are plotted as filled bars on the left and MFI of bright sorted clones as open bars on the right.

11.1. Clonal brightness

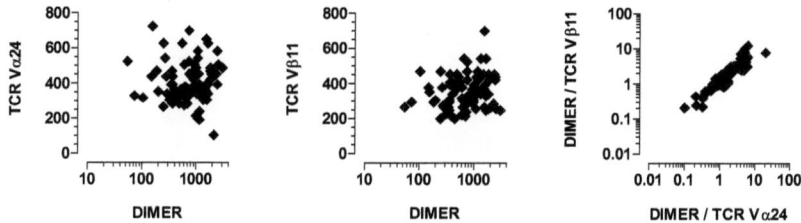

Figure 11.6
Correlation of the TCR alpha- and beta-chain with the DIMER. The MFI values of DIMER, TCR Vα24, TCR Vβ11 and quotients thereof are plotted for all BG clones. P values for (Spearman nonparametric) correlation analyses are not significant except P<0.0001 for the graph on the right.

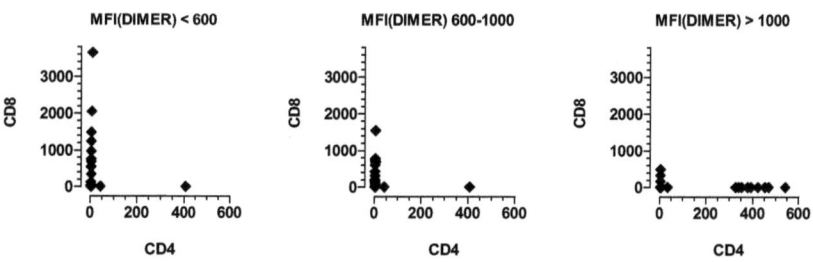

Figure 11.7
Relation of the classical CD4 and CD8 co-receptors to the DIMER. CD4 is plotted versus CD8 for three subdivisions of DIMER MFI, namely clones staining below 600, from 600 to 1000 and over 1000 (dot plots from left to right).

Neither CD4 nor CD8 do interact or partner with CD1d, at least not around the epitopes of the mAb used, as no inhibition of DIMER was observed (data not shown). Because iNKT cells, as implied by their name, express natural killer (NK) markers like natural cytotoxicity receptors (NCR), it was checked for their presence and the expression of killer cell immunoglobulin-like receptors (KIR) as well as for the surface levels of killer cell lectin-like receptors (KLR). Examining the sorted lines showed that a minor percentage of total cells was positive for the NCR NKp30, NKp44 and NKp46 (0.88%, 0.56% and 0.97% of the dull line versus 0.22%, 0.40% and 0.19% of the bright line respectively). The KLR NKG2D, one major activatory receptor on NK cells, was more widely expressed with 29.13% of the dull- and 23.35% of the bright-sorted cells being positive. Further investigations, including CD94 and CD161, on the lines and on the BG clones did not reveal significant association of either or both markers with dull or bright DIMER staining (data not shown). Altogether these data led us to revise the hypothesis from 'dull and bright sees the antigen the same way and the staining difference is due to a co-receptor' to the hypothesis 'dull and bright sees the antigen in different ways due to differences in (the variable and joining regions of) the β-chain'. A hint to confirm the latter hypothesis came from DIMER blocking experiments performed on the dullest and on the brightest BG clone. Blocking by a anti-TCR Vα24 required two times more mAb for the brightest compared to the dullest clone, whereas five times or six times more mAb was necessary for anti-TCR Vβ21 or anti-CD1d mAb (see Table 11.2, p.68), respectively.

Table 11.2 DIMER blocking - 50% blocking concentration (µg/ml)

mAb	epitope on	BGA1	BGA89	fold difference
C15	Vα24	0.095	0.18	1.9
C21	Vβ11	0.450	2.15	4.8
42	CD1d	3.250	20.00	6.2

Conclusively, different variable and joining regions of the TCR β-chain may make up antigen specificity even though the participation of a to date unknown co-receptor can not be formally ruled out. Investigations to correlate DIMER staining with cytokine profiles or release or activation thresholds were not conclusive so far; even taking into account the surface molecule fingerprints of the iNKT cells. If iNKT cells predominantly play their innate or adaptive skills might be depending on the type of APC and/or the environment where they become activated in and has to be investigated in future.

More recent findings revealed the αGC-CD1d structure [Koch et al. 2005], the structure and binding kinetics of three different human αGC-CD1d-specific TCR [Gadola et al. 2006 b] and the influence on the affinity of iNKT TCR and the threshold of iNKT cell activation by the length of lipids bound to human CD1d molecules [McCarthy et al. 2007]. The architecture of CD1 proteins has been reviewed [Zajonc & Wilson 2007] and the crystal structures of lipid-CD1 complexes disclose differences in the mode of presentation of lipids by the CD1 isoforms. Whereas group 1, especially CD1a and CD1b, anchors the lipid backbone inside the hydrophobic binding grooves ('lipid anchoring'), group 2 CD1d positions the polar ligand headgroup(s) through a precise hydrogen-bonding network in well-defined orientation at the T cell recognition interface ('headgroup positioning'). Even more important is the occurrence of small conformational changes in the CD1d protein upon binding of αGC due to increased polar interactions with the α1 and α2 helices that line the antigen binding groove [Silk et al. 2008 b]. Additional fragmentary electron density data indicates that the binding groove of non-lipid-bound CD1d may be partially occupied by some small-molecule species to overcome the inherent lability of CD1d molecules. What becomes readily visible in the absence of a crystal lattice, where CD1d molecules lacking bound αGC are unstable. Comparison of the filled and empty CD1d showed several structural disparities arising primarily from alterations in side-chain conformations to fill vacant space [Koch et al. 2005]. These modelings provide the basis for a scheme of ligand-induced changes in the APM (performed on human CD1d and αGC and some analogues). The model predicts how the length of the lipid tail(s), in particular inside the F' cavity, propagates structural changes to the TCR recognition interface [McCarthy et al. 2007]. Furthermore, that might explain how nonglycosidic CD1d lipid ligands activate human and murine iNKT cells [Silk et al. 2008 a] without engaging the hydrogen-bonding network that is formed between the CD1d α helices and the headgroups of e.g. glycosidic antigens.

Although earlier studies tried to explain the interaction of iNKT TCR to antigen-loaded CD1d by iNKT TCR sequence analysis [Kawano et al. 1999] or site-directed mutagenesis of CD1d [Burdin et al. 2000], conclusive and surprising insights were only reached by the groundbreaking iNKT TCR αGC-CD1d co-crystal [Borg et al. 2007] and an iNKT TCR αGC-CD1d contact interaction analysis [Wun et al. 2008]. Most earlier studies focused on the complementarity determining region (CDR)3 of the TCR β-chain as it is supposed to contain the most diversity due to the fact that it is located in the V-D-J joint [Danska et al. 1990; Saada et al. 2007]. Alanine substitution analysis revealed that the Tyr48 and Tyr50 located in the CDR2β contributed the most in the interaction of the TCR β-chain by binding to the α1 helix of CD1d (in addition to Asn53 and Glu56) but only one residue, Tyr103, in the CDR3β chain negligibly added to the interaction by binding to the α2 helix [Wun et al. 2008]. The CDR2β loop is positioned directly over the F' pocket in the CD1d as seen in the co-crystal [Borg et al. 2007]. That may explain how the TCR can sense different lipid tail lengths through structural propagation by F' pocket occupation [McCarthy et al. 2007]. From the inverse point of view, that is in line with our findings that a series of more than eighty iNKT cell clones are able to demonstrate an amazing range of DIMER MFI for the very same antigen-CD1d complex, namely CD1d loaded by αGC (see Figure 11.5, p.66). These structural data have already been successfully integrated to design functionally enhanced lipid activators of iNKT cells and will continue to be integrated for design of other chemotherapeutic agents or immunostimulatory compounds for a variety of immune-mediated diseases.

One major difference is the role of the co-receptors CD4 and CD8 in peptide-MHC versus lipid-CD1 multimer systems. As the importance of CD4 and CD8 is well documented for peptide-MHC interactions with TCR [Mallaun et al. 2008; Palmer & Naeher 2009; Rudolph et al. 2006; Wang & Reinherz 2002], a co-receptor for CD1 molecules has not been found to date and CD1-reactive T cells have an unrestrained pattern of co-receptor expression comprising CD4 or CD8 or neither one. Because for CD1-restricted TCR the major interaction partner is the CD1 molecule and to a marginal part the lipid antigen [Borg et al. 2007; Wun et al. 2008], secondary factors as temperature can have even more dramatic effects on staining than those reported for MHC-restricted TCR [Whelan et al. 1999]. Predictions how temperature will affect all systems remain unreliable, therefore it is advisable to control the effects of temperature on each individual system.

Tetramers of peptide-MHC, which are thought to engage three different TCR, are more than sufficient for most staining applications [Wooldridge et al. 2009]; in many cases, a simple peptide-MHC dimer is effective. Conclusively, multimerization of the DIMER on dextran does not abolish the property of the DIMER to sense the affinity of the a clonal TCR to αGC but tremendously increases the avidity (data not shown). High molecular weight dextran polymers have an osmotic pressure effect during the staining procedure when used at high concentrations;

that can easily be circumvented or at least ameliorated by dilution. Avoiding an osmotic pressure effect is of importance when staining PBMC, as shifts in the forward and side scatter profile would complicate the identification of desired populations. Further perspectives include the establishment of an enhanced multimer system tested for a wide range of lipid antigens that will be of great value as diagnostic tool to detect rare lipid-specific and CD1d-restricted T cell populations in patients with different types of diseases.

Tetramers are as well used to examine the activation requirements of T cells. However, the possibility of antigen transfer from soluble antigen-APM complexes to APM molecules on the surface of the T cells themselves, or other cells in the assay, can not be ignored. Therefore systems are developed to exclude at least influences from third party cells other than T cells. One prerequisite was the realization that cross-linking cell surface TCR and CD8 with soluble peptide-MHC class I tetrameric complexes produces a pattern of early tyrosine phosphorylation that resembles the one induced by antigen-pulsed APC [Purbhoo et al. 2001]. Moreover, co-stimulation signals from other molecules such as CD28 are not obligatory for T cell activation by tetramers. Cross-linking with anti-CD3 is known to produce a different intracellular signaling pattern. Even if CD8 is not required for tetramer binding, experiments in mice have confirmed the necessity of CD8 binding for tetramer-induced activation.

Tetramer activation of T cells can be attained at over 1000-fold lower concentrations of tetramer than required to visualize even small shifts in MFI in FACS experiments [Wooldridge et al. 2009]. The remarkable sensitivity of antigen-APM multimer-induced T cell activation is in turn demonstrated by such faint shifts that may correspond to only a few engaged tetramers per T cell. In addition to trigger a normal pattern of T cell signaling cascades, tetramer activation results in lytic granule as well as a full range of cytokine and chemokine release and the induction of various cell surface activation markers.

After the establishment of DIMER staining by FACS, we were also interested in the iNKT cell response upon antigen encounter. Most prominent might be the distinction of type 1 helper T cells (T_H1) and type 2 helper T cells (T_H2), the type preponderantly induced by the immune system is in many cases critically important for the outcome of a disease or infection. As iNKT cells unlike classical $\alpha\beta$ T cells present an activated phenotype and do not require any priming to release large amounts of cytokines, further confirmed by data of tetramer-activated classical T cells, we thought of using the DIMER immobilized on plastic and loaded with the antigen of interest as stimulant.

After successful establishment of a standard protocol to activate iNKT cells by plate-bound CD1d loaded with αGC using the same engineered CD1d molecules as in DIMER staining, diverse chemically synthesized compounds were investigated in order to find new lipids stimulating iNKT cells. Thanks to a fruitful collaboration with a group of chemists, a whole series of compounds was accessible that were designed to test the role of the antigen headgroup and acyl chains in iNKT cell activation. These studies were published [Mathew et al. 2009 a; Mathew et al. 2009 b,c; Rajan et al. 2009] and their biological aspects are summarized in the subsequent four chapters. Before assessing several structural features of iNKT cell-ligands, antigenicity of a new αGC analogue was examined.

11.2. N-acetyl-2-amino-2-deoxy-α-galactosyl 1-thio-7-oxaceramide, a new α-galactosylceramide analogue

The N-acetyl-2-amino-2-deoxy-α-D-galactopyranosyl 1-thio-7-oxaceramide (see Table 11.3 (**1**), p.71) was synthesized [Rajan et al. 2009] by substituting a 7-oxasphingosine triflate with α-D-N-acetyl-1-thiogalactosamine. The triflate was obtained from an azide and the thiol was prepared according to a known procedure from α-D-galactosamine hydrochloride. As compared to ceramide, **1** is neither a substrate of ceramide kinase, consistent with the absence of the C(1)-OH group, nor an inhibitor of ceramide phosphorylation by ceramide kinase. While it partially displaced CD1d-bound lipids, it failed to stimulate iNKT cells when presented by human CD1d-transfected cells. These results suggest that it binds weakly to recombinant CD1d, but does not form immunogenic complexes with CD1d.

Table 11.3 Structure of a new galactosaminide αGC analogue

Number	Name	Structure
1	N-acetyl-2-amino-2-deoxy-α-D-galactopyranosyl 1-thio-7-oxaceramide	

Several α-D-galactopyranosyl ceramides [Brodesser et al. 2003; Radin 2003; Shimamura et al. 2007; Tan & Chen 2003; Vankar & Schmidt 2000], especially KRN-7000 [Miyamoto et al. 2001; Morita et al. 1995 b], OCH [Franck & Tsuji 2006; Murata et al. 2005; Ndonye et al. 2005], and RCAI-61 [i] [Tashiro et al. 2008], incorporating phytosphingosine or sphingosine moieties, possess important biological properties [Bendelac et al. 2007; Godfrey & Berzins 2007; Kronenberg 2005; Tupin et al. 2007], comprising immunostimulating [Barbieri et al. 2004; Costantino et al. 2002; Yang et al. 2004] and antitumor activities [Bleicher & Cabot 2002; Cuvillier 2002; Gulbins & Grassme 2002; Kawase et al. 2002; Maceyka et al. 2002; Obeid et al. 1993; Pettus et al. 2002; Shikata et al. 2003]. Several analogues of these α-D-galactopyranosides were synthesized, and they also display notable biological activities [Franck & Tsuji 2006; Yang et al. 2004] (see [Rajan et al. 2009] for further references). The synthesis and evaluation of a thioglycoside analogue of KRN7000 has recently been published [Dere & Zhu 2008], prompting us to report our results on the synthesis and evaluation of **1**. Replacing the galactosyl by an N-acetyl-2-amino-2-deoxygalactosyl moiety was thought to contribute to elucidating the effect of C(2)-OH of the glycon on TCR recognition, while the S–glycosyl bond will confer stability against chemical and enzymatic cleavage [Driguez 2001]. 7-oxasphingosines behave similarly to ceramides, and allow to independently modify the headgroup and lipid moiety of sphingosine, an advantage that has since been realized using olefin metathesis [Nussbaumer 2008; Yamamoto et al. 2006]. Conceivably, **1** may be obtained either by glycosidation [Pachamuthu & Schmidt 2006] of a 1-thioceramide or of a 1-thiosphingosine, or by nucleophilic substitution by an N-acetyl α-D-galactopyr-anosyl thiol of a ceramide possessing a leaving group at C(1) and preferentially a non-participating neighboring group, such as an N_3 substituent [Herzner et al. 2000]. We opted for the second strategy, assuming that nucleophilic substitution will prevail over single electron transfer from the thiolate to

---
[i] OCH is a truncated and RCAI-61 the 6'-O-methyl analogue of KRN7000

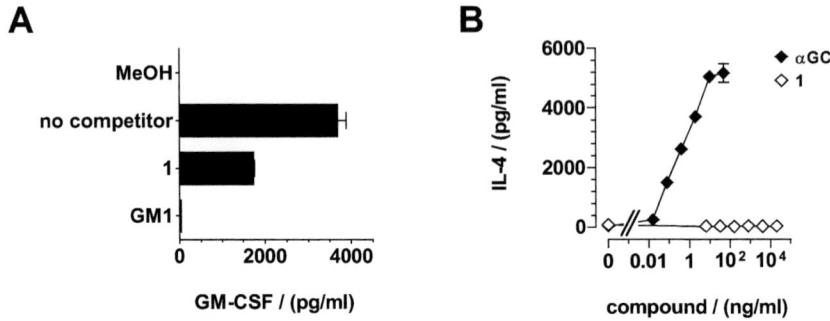

Figure 11.8
Binding to CD1d and failure to activate iNKT cells. (A) N-acetyl-2-amino-2-deoxy-α-D-galactopyranosyl 1-thio-7-oxaceramide (**1**) was tested for its capacity to prevent binding of αGC to plate-bound CD1d and subsequent T cell response to αGC. Supernatants were taken and released GM-CSF and IL-4 (data not shown) were measured by ELISA and expressed as pg/ml ± SD of triplicates. Weak but significant competition of **1** with αGC was seen as compared to complete inhibition of αGC by ganglioside monosialic acid (GM1) used as control competitor. (B) In contrast, **1** (open diamonds) failed to activate iNKT cells when titrated on CD1d transfectants and compared to αGC (closed diamonds). Supernatants were assessed for release of IL-4, IFN-γ and TNF-α giving similar results.

the N_3 group, and reported on the synthesis of **1** and the results of its evaluation as substrate of ceramide kinase [Spiegel & Milstien 2007] and as T cell-stimulatory antigen [Staros et al. 1978].

The galactosaminide **1** is neither cytotoxic to APC nor to iNKT cells at the tested dosage (up to 20 μg/ml), as assessed by flow cytometry (data not shown). First, binding to human CD1d and displacement of the strong agonist αGC was investigated. CD1d was attached to the plastic and incubated with **1** at molar excess over αGC, first with the former and then with the latter one. **1** partially reduced the activation of iNKT cells by αGC (see Figure 11.8 (A), p.72), indicating a weak capacity to prevent binding of αGC to CD1d under these experimental conditions. Next, **1** was investigated for stimulation of iNKT cells when presented by living APC. Different APC were used to exclude cell type-specific effects. In all experiments, there was no activation of iNKT cells (see Figure 11.8 (B), p.72). Lack of iNKT cell stimulatory potential may be ascribed to the replacement of the OH group at C(2) of the glycon by the acetamido group. Replacement of OH at C(2) of the glycon with an NH_2 group also abolishes iNKT cell activation [Miyamoto et al. 2001]. The OH group at C(2) establishes H-bonding with two amino acids (R95 and G96) on the α-chain of iNKT TCR [Borg et al. 2007] that is mandatory for T cell activation, as shown by the CD1d:αGC binding footprint for the NKT TCR [Wun et al. 2008]. The observed weak displacement capacity may be attributed to the different constitution and rigidity of the oxasphingosine moiety as compared to the phytosphingosine moiety of αGC. The different structure of oxasphingosine might impair the interaction between **1** and the CD1d F' pocket, and influence the position of the sugar head, preventing an optimal interaction with the TCR.

11.3. Sphingolipid analogues based on 7-oxasphingosine and 7-oxaceramide

Table 11.4 New sphingolipid analogues based on 7-oxasphingosine/-oxaceramide

7-oxasphingosine/-oxaceramide analogues with a modified C(1)-OH or amide moiety

The analogues **7–9** [i] of 7-oxaceramide and 7-oxasphingosine were synthesized from the known azidosphingosine **21**. The 1,4-disubstituted 1,2,3-triazole analogues **10–16** of ceramides were synthesized by the click reaction of the known azide **24** (see Table 11.4, p.73). None of the analogues **7–15** was active as inhibitor of sphingosine kinase type 1 and of acid sphingomyelinase, whereas **16** is a weak inhibitor of sphingosine kinase type 1. Triazoles **10**, **11**, and **15** did not inhibit ceramide phosphorylation by ceramide kinase, and none of **7**, **8**, and **10–15** activated iNKT cell clones when presented by human CD1d-transfected APC or by plate-bound human CD1d [Rovina et al. 2006]. Triazoles **14** and **15** prevent binding of αGC to plate-bound human CD1d and subsequent T cell response to αGC. Only **15** reduced activation by αGC significantly and independently of the cytokine measured (see Figure 11.9, p.75).

The ability of sphingolipids and glycosphingolipids to modulate apoptosis [Brodesser et al. 2003; Cuvillier 2002; Gulbins & Grassme 2002; Maceyka et al. 2002; Obeid et al. 1993; Padron 2006; Shikata et al. 2003] and immune responses [Barbieri et al. 2004; Costantino et al. 2002; Pettus et al. 2002; Plettenburg et al. 2002; Wu et al. 2008; Yang et al. 2004] is likely to depend on specific receptor interactions with both the polar headgroup and the lipid tails of the sphingosine and fatty acid moieties. The exploration of these interactions is based on crystal-structure analyses of receptors in complex with the lipids [Koch et al. 2005; Snook et al. 2006], and on the synthesis and biological evaluation of analogues [Delgado et al. 2006; Julina et al. 2004; Li et al. 2002].

1,2,3-Triazoles are non-hydrolyzable bioisosters of amides [Brik et al. 2005; Kolb & Sharpless 2003], readily obtained by the click reaction [Appendino et al. 2007; Bock et al. 2006; Brik et al. 2005; Finn et al. 2008; Kolb & Sharpless 2003; Moses & Moorhouse 2007; Rostovtsev et al.

---

[i] Note that the compound numbers in this section refer to [Mathew et al. 2009 b]

2002]. They mimic geometric and electronic features of a configurationally biased amide bond, and participate in dipole–dipole interactions [Appendino et al. 2007; Gajewski et al. 2007] and H-bonding. They may act as H-bond acceptors and as (weak) donors, depending on their substitution. In 1,4-disubstituted 1,2,3-triazoles, N(2) and N(3) exhibit H-bond accepting properties, whereas the strong dipole moment of triazole causes the polarization of H-C(5) that may act as a weak H-bond donor [Horne et al. 2004; Palmer et al. 1974; Purcell & Singer 1967] (see [Mathew et al. 2009 *b*] for further references). The triazole ring of the envisaged analogues **10**–**16** [Mathew et al. 2009 *b*] orients the acyl chain in a way that prevents a parallel orientation with the alkyl chain of the sphingosine moiety, and its polar substituents may serve as H-bond acceptors in a position differing from the C=O group of the C(2)-acylamino chain. These features are of interest, as the crystal structure of αGC bound to the human and mouse CD1d receptor shows that the two alkyl chains diverge from each other [Koch et al. 2005], suggesting that the configuration of the amide moiety, besides its H-bond-acceptor and -donor properties, plays an important role in interacting with receptors. The triazoles should be readily available from the known azide intermediate in the synthesis of 7-oxaceramide [Rajan et al. 2004]. After the synthesis of the triazoles **10**–**16** was completed, a synthesis of related ceramides [Kim et al. 2007] and O-glycosides [Lee et al. 2007] that stimulated cytokine production, comparable to the one of αGC, and exhibited a stronger T_H2 cytokine response was published. Biological properties for which we considered testing these analogues comprise T cell suppression [Savage et al. 2006], and the interaction with sphingosine and ceramide kinases, and with acid sphingomyelinase [Baumruker & Prieschl 2002].

Compounds **7**, **8**, and **10**–**15** did neither show any activatory capacity on iNKT cell clones when presented by human CD1d-transfected APC nor by plate-bound human CD1d (data not shown). None of the compounds proved cytotoxic for APC or iNKT cells at tested doses (up to 20 µg/ml), as assessed by flow cytometry (data not shown). Lack of stimulatory activity may not exclude the capacity of these compounds to bind to CD1d. This was investigated using a competition assay to evaluate the capacity of the compounds to prevent binding of αGC to CD1d and, thus, to prevent T cell activation. Only compound **15** was able to compete, although this competition was only achieved at 15-fold molar excess over αGC (see Figure 11.9, p.75). In additional experiments, we investigated whether **15** was capable of competing with αGC in living cells. In contrast to the strong inhibition seen in the plate-bound assay, **15** was incapable of reducing T cell activation at all tested doses of αGC (data not shown).

This unexpected discrepancy between plate-bound and living cell competition assays might be explained with the instability of the thioamide moiety of **15** inside living cells. To test the stability of the thioamide **15**, we performed plate-bound assays at pH 4 and 7. At pH 4, an increase in cytokine release as compared to pH 7 was seen, and a partial degradation of **15** was thus assumed (see Figure 11.10, p.75). In living cells, **15** requires trafficking to lysosomes, where lipid antigens are loaded onto CD1d molecules [Jayawardena-Wolf et al. 2001]. Therefore, it is tempting to speculate that the acidic late-endosomal microenvironment affects the stability of **15**.

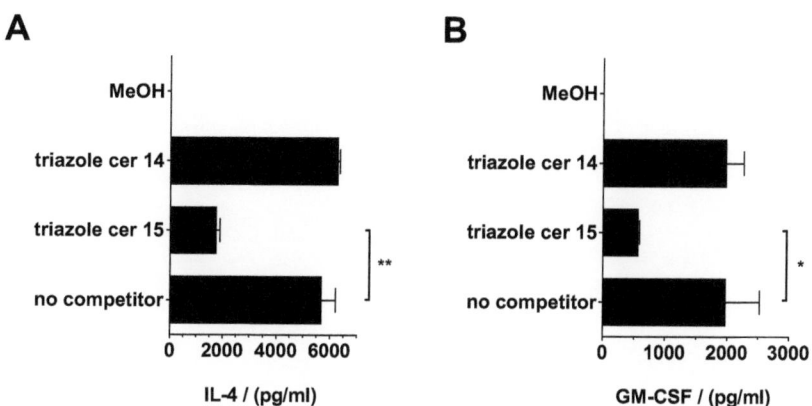

Figure 11.9
Competition of triazoles with alpha-galactosylceramide. Ceramide analogues **14** and **15** were tested for their capacity to prevent binding of αGC to plate-bound human CD1d and subsequent T cell response to αGC. Supernatants were taken and released IL-4 (A) and GM-CSF (B) were measured by ELISA and expressed as pg/ml ± SD of duplicates. Only **15** was significantly reducing activation by αGC independently of the cytokine measured (**: $P<0.01$; *: $P<0.05$).

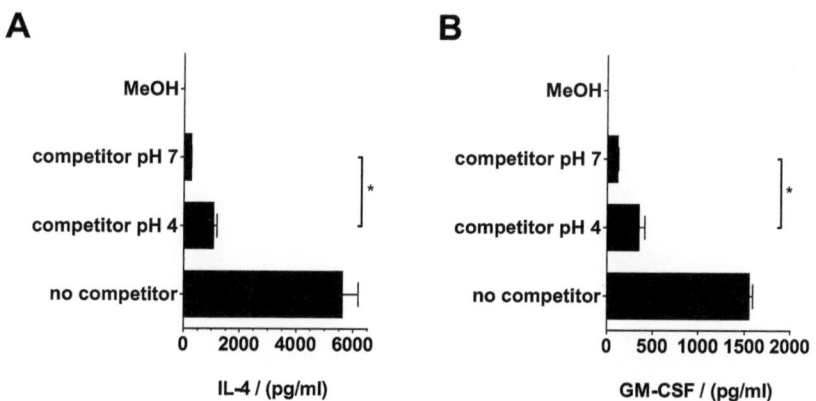

Figure 11.10
A triazole with sensitivity to low pH. A competition experiment was performed at pH 4 or 7. Supernatants were taken and released IL-4 (A) and GM-CSF (B) were measured by ELISA, and expressed as pg/ml ± SD of duplicates. At pH 4, triazole **15** shows less competition than at pH 7, independently of the cytokine measured (*: $P<0.05$).

11.4. 7-aza- and 7-thiasphingosines

Table 11.5 Structures of 7-aza- and 7-thiasphingosines

The trifluoroacetamide **24**, the 7-thiasphingosine **5**, the sulfone **6** and the 7-azasphingosine **7**

The synthesis of 7-oxasphingosine (**3** [i]) and 7-oxaceramide (**4**) was improved [Mathew et al. 2009 b] by starting from the 4-methoxybenzyl-protected D-galactal **9**. The sphingosine analogues **5**–**7** and **24** were synthesized via the azido alcohol **13** (see Table 11.5, p.76). The 7-thiasphingosine **5** is a poorer substrate for both isoforms of sphingosine kinase than sphingosine, but showed a slight preference for sphingosine kinase type 2. The sulfone **6** and the 7-aza compounds **7** and **24** were not phosphorylated by either type 1 or type 2, and none of **5**–**7** and **24** activated iNKT cell clones when presented by CD1d-transfected APC or by plate-bound CD1d. Only **7** and **24** associated with plastic-bound recombinant CD1d and prevented stimulation of iNKT cells by αGC in plate-bound activation assays (see Figure 11.11, p.77).

Sphingolipids are structural components of the cell and exert multiple functions in cell signaling, including the role of D-erythro-sphingosine and ceramide in the regulation of apoptosis [Brodesser et al. 2003; Cuvillier 2002; Gulbins & Grassme 2002; Maceyka et al. 2002; Obeid et al. 1993; Pettus et al. 2002; Shikata et al. 2003], and the role of sphingosine-1-phosphate and ceramide in immune responses [Barbieri et al. 2004; Costantino et al. 2002; Padron 2006; Plettenburg et al. 2002; Wu et al. 2008; Yang et al. 2004]. These functions are thought to depend on specific interactions between receptors, and both the lipid tail (two tails for ceramide) and the polar headgroup of sphingosine and ceramide [Delgado et al. 2006]. We began exploring these interactions by synthesizing and evaluating analogues, and so far reported on modifications of the polar headgroup and on the replacement of the C(7)H$_2$ group by an O-atom, as in **3** and **4**, a replacement that did not appear to affect the tested biological properties [Rajan et al. 2004]. This may be different for analogues where the C(7)H$_2$ group is replaced by bulkier, more polar, or charged groups, and we considered the 7-thia and 7-aza analogues **5**–**7** of interest.

We now describe the modified synthesis of 7-oxasphingosine and 7-oxaceramide, the synthesis of 7-thia- and 7-azasphingosines, and the evaluation of these analogues as substrates of sphingosine kinases, as CD1d ligands, and as activators of iNKT cells. The results clearly show that the electronic environment and/or nature of the groups near C(7) of sphingosine do have an effect

---

[i] Note that the compound numbers in this section refer to [Mathew et al. 2009 c]

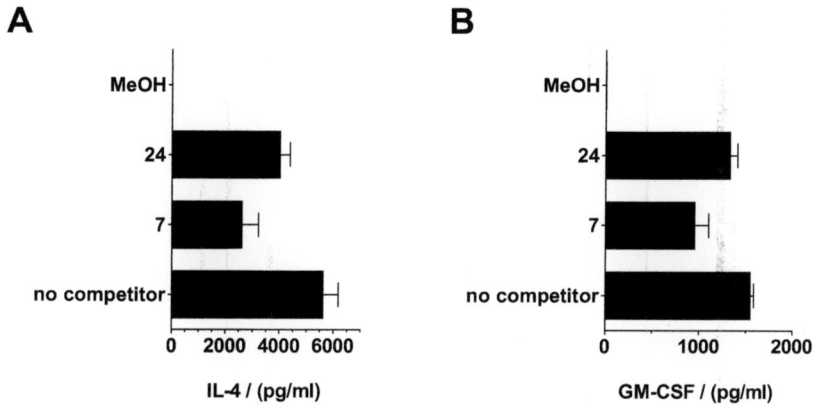

Figure 11.11
Competition of azasphingosines with alpha-galactosylceramide on plastic CD1d. The 7-azasphingosines **7** and **24** are able to compete with αGC for plate-bound CD1d. Weak reduction of both (A) IL-4 and (B) GM-CSF release is observed.

on the biological activity.

Compounds **5–7** and **24** did neither activate iNKT cell clones when presented by human CD1d-transfected APC nor by plate-bound human CD1d (data not shown). Compounds **5**, **7**, and **24** were cytotoxic for APC and iNKT cells above 5 µg/ml as assessed by flow cytometry (data not shown), and, therefore, cytotoxicity could mask their stimulatory capacity. Next, binding to CD1d was evaluated in cell-free competition assays. When T cells were stimulated with plate-bound recombinant CD1d and αGC, only the 7-azasphingosines **7** and **24** were partially inhibitory at a 20-fold molar excess (see Figure 11.11, p.77). The large molar excess needed to compete with αGC could be attributed to the fact that these lipids have only one tail, and thus bind weakly to CD1d, whereas αGC binds with high affinity because it has two lipid tails. The compounds were also tested for inhibition of the response to αGC using living APC. This is a more sensitive assay than the one using plate-bound CD1d, but could be affected by other ways of lipid-influence on APC. All compounds were carefully applied at nontoxic doses. As observed with plate-bound assays, also with living APC, compounds **7** and **24** slightly inhibited T cell response (see Figure 11.12, p.78). This was observed with all three tested cytokines (IL-4, IFN-γ and TNF-α).

In conclusion, both 7-azasphingosines **7** and **24** do not stimulate iNKT cells but bind to human CD1d, and might be used to influence iNKT cell responses.

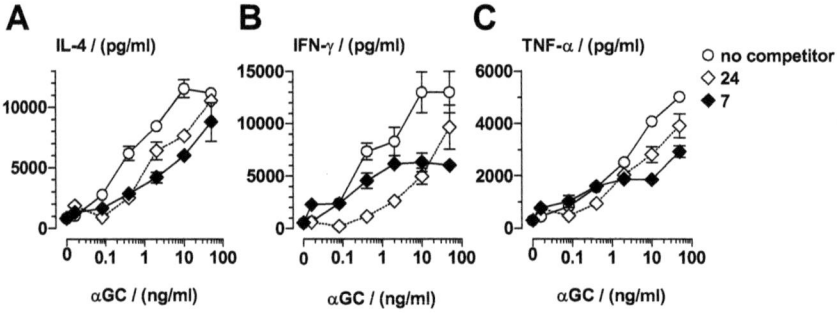

Figure 11.12
Competition of azasphingosines with alpha-galactosylceramide on APC. The 7-azasphingosines **7** and **24** are able to compete with αGC for CD1d on living APC. Reduction of iNKT cell activation is seen by release of (A) IL-4, (B) IFN-γ and (C) TNF-α. Compounds **7** and **24** affect potency and/or efficacy of αGC differently regarding the cytokine measured (open circles = no competitor, open diamonds = **24**, closed diamonds = **7**).

11.5. 4,5,6-tri-substituted piperidinones as conformationally restricted ceramide analogues

Table 11.6   Structures of piperidinones with a single lipid tail

The 4,5,6-tri-substituted piperidinones without an additional alkyl chain at C(3)

7	8	15	16

| 17 X = NH$_2$ | 18 X = NH$_2$ | 19 X = NH$_2$ | 20 X = NH$_2$ |
| 21 X = NMe$_2$ | 22 X = NMe$_2$ | 23 X = NMe$_2$ | 24 X = NMe$_2$ |

The conformationally based piperidinone sphingosine analogues **7**, **8**, **15**, and **16** [i] were synthesized [Mathew et al. 2009 a] from allylic alcohol **34** via lactams **31** and **32**. The L-*arabino* diol **7** and the L-*ribo* diol **8** were transformed into the amino alcohols **17**–**24** (see Table 11.6, p.79). The L-*gluco* ceramide analogues **43**, **46a**, and **47**, and the L-altro ceramide analogues **51a** and **52** (see Table 11.7, p.80) were synthesized from either **31**, or **32**. The L-*ribo* diols **8** and **16**, and amino alcohols **19** and **20** inhibit sphingosine kinase 1, while the L-*arabino* analogues **7**, **15**, **17**, and **18** are inactive. The L-*arabino* and the L-*ribo* dimethylamines **21**–**24**, the L-*gluco* ceramide analogues **43**, **46a**, and **47**, and the L-altro ceramide analogues **51a** and **52** did not block sphingosine kinase 1. Neither the L-*arabino* diol **7** nor the L-*ribo* diol **8** inhibited sphingosine kinase 2 or ceramide kinase. The L-*arabino* diols **7** and **15** stimulate iNKT cells when presented by living APC and also by plate-bound human CD1d, whereas the L-*ribo* diols **8** and **16**, the L-*arabino* amino alcohols **17**–**18**, and the dimethylamines **21**–**22** did not activate iNKT cells. The L-*gluco* ceramide analogues **43**, **46a**, and **47** had a strongly stimulatory effect on iNKT cells when presented by living APC and also by plate-bound CD1d, whereas the L-altro ceramide analogue **52** activated only weakly. All activatory compounds preferentially induce the release of proinflammatory cytokines, indicating the formation of a stable lipid-CD1d-TCR complex.

---

[i] Note that the compound numbers in this section refer to [Mathew et al. 2009 a]

Table 11.7 Structures of piperidinones with two lipid tails

The 4,5,6-tri-substituted piperidinones with an additional alkyl chain at C(3)

(Glyco)sphingolipids and more specifically ceramides adopt a parallel orientation of the lipid chains in the cell membrane, requiring a (Z)-configuration of the amide moiety, while the (E)-configuration of amides is preferred in solution and in the solid state by a free energy difference of ca. 1.2 kcal/mol [Langley et al. 2005; Pawar et al. 1998]. There is no a priori reason why the amide moiety of a ceramide complexed to a receptor should either adopt the (E)- or the (Z)-configuration. The crystal structure of αGC bound to the human and the mouse CD1d receptor shows the (E)-configuration, i.e., the two alkyl chains diverge from each other [Koch et al. 2005], suggesting that the configuration of the amide moiety and not only its hydrogen bond acceptor and donor properties may be crucial for the interaction with receptors. Yet, only very few compounds have been prepared that may be considered conformationally well-defined ceramide analogues and are biologically active. The isoquinolines **1** were investigated as conformationally restricted analogues of ceramide, and act as ligands of protein phosphatase 2A, a ceramide-binding protein that has been implicated in signal transmission [Leoni et al. 1998]. The uracil and thiouracil derivatives **2a** and **2b**, respectively, are further examples, exhibiting moderate anti-tumor activity and toxicity *in vitro* and *in vivo* [Macchia et al. 2001]. Finally, the analogues **3** and **4** [Brodesser et al. 2003] inhibit GM-2 synthase. In 2006, AWA16-1 awajanomycin (**5**) and its derivative **6**, a piperidinone based cytotoxic agent, was isolated from the marine fungus *Acremonium* sp. [Jang et al. 2006]. Both **5** and **6** inhibit the growth of A549 cells with half maximal inhibitory concentration (IC_{50}) values of 27.5 and 46.4 µg/ml, respectively.

We were interested in restricting the conformation of the headgroup of ceramide, and in particular of the amide moiety, as they are recognized and modified by sphingosine- and ceramide-metabolizing enzymes. The conformationally restricted L-*arabino* and the L-*ribo* piperidinones [i] **7** and **8**, respectively, may be considered analogues of sphingosine, and the L-*gluco*, L-*manno*, L-*altro*, and L-*allo* piperidinones **9**–**12**, respectively, may be considered ceramide analogues mimicking the (Z)-configuration of the amide. The piperidinones **9**–**12**, substituted at C(3) and C(6), possess most of the structural features that are required for the biological functions of ceramides, *viz.*, C(1) and C(3) hydroxy groups [Koch et al. 2005], the C(2) *N*-acylamino substituent, a trans

---

[i] Piperidinones are also known as hexonolactams. The configuration is specified by the carbohydrate prefix in the *Theoretical Part*, and by the (R/S)-designation in the *Experimental Part* [Mathew et al. 2009 a].

11.5. 4,5,6-tri-substituted piperidinones as conformationally restricted ceramide analogues

double bond in the lipid chain of the sphingosine moiety [Koch et al. 2005; Malewicz et al. 2005], and a substituent at C(3) corresponding to the lipid part of the N-acyl substituent. As compared to sphingosines and ceramides, the piperidinones possess an additional center of chirality, resulting in four diastereoisomeric pairs of enantiomers of the sphingosine analogues **7** and **8**, and in eight diastereoisomeric pairs of enantiomers of the ceramide analogues **9-12**. Two and four diastereoisomers possess the same configuration at the corresponding centers as D-erythrosphingosine (**13**) and ceramide (**14**), respectively. Starting from D-galactose will result in these two and four diastereoisomers **7–12**. Biological properties for which the piperidinone analogues should be tested include the inhibition of sphingosine [Spiegel & Milstien 2007] and ceramide kinases [Thevissen et al. 2006] and the effect on the production of lymphokines [Baumruker & Prieschl 2002].

iNKT cells recognize a variety of lipid antigens presented by the CD1d molecule. Therefore, we tested whether piperidinones are capable of stimulating this population of human T lymphocytes. The piperidinones were first tested for cytotoxic effects *in vitro*. T cells were incubated overnight with increasing doses of the sonicated piperidinones and the next day, cell death was assessed by measuring the uptake of PI or 7AAD using a CYAN™ADP flow cytometer (Beckman Coulter). The LC_{50} were calculated for each compound (see Table 11.8, p.81).

Table 11.8 Median lethal concentrations (LC_{50}) of piperidinones

Compound	Configuration	LC_{50} [i]	Compound	Configuration	LC_{50}
7	L-*arabino*	8	8	L-*ribo*	nd [ii]
15	L-*arabino*	9	16	L-*ribo*	nd
17	L-*arabino*	9	19	L-*ribo*	4.5
18	L-*arabino*	4.5	20	L-*ribo*	4
21	L-*arabino*	8	23	L-*ribo*	9.5
22	L-*arabino*	8	24	L-*ribo*	nd
43	L-*gluco*	>20	51a	L-*altro*	nd
46a	L-*gluco*	>20 [iii]	52	L-*altro*	7
47	L-*gluco*	>20			

[i] Given in µg/ml
[ii] Not determined
[iii] In activation it seemed toxic above 5 µg/ml (see Figure 11.15 (A-C), p.85)

The listed piperidinones (see Table 11.8, p.81), with the exception of L-*altro* **51a**, were used to stimulate iNKT cells (see Figure 11.13, p.83) presented by THP1 cells transfected with human CD1d gene (THP1-hCD1d). In parallel experiments, plate-bound recombinant soluble human CD1d was loaded with the relevant lipids and used to stimulate iNKT cells (see Figure 11.14, p.84). Briefly, THP1-hCD1d cells or soluble plate-bound recombinant human CD1d (5 µg/ml) were incubated with the sonicated compounds at the indicated concentrations in the presence of T cells. For competition assays a fixed dose of the piperidinones was given 4.5 h in advance of titrating αGC and addition of T cells. Culture supernatants were collected after 24 h, and iNKT cell released cytokines were measured by ELISA [Shamshiev et al. 2002]. The L-*arabino* diols **7** and **15** activated iNKT cells (see Figure 11.13, p.83), although they possess only one lipid tail. This finding was unexpected, since all the iNKT cell stimulatory compounds described in the literature at the time possess two lipid chains. More recently, several studies reported the binding of lipids with a single tail to CD1d [Cox et al. 2009; Yuan et al. 2009] and their activatory capacity for iNKT cells [Fox et al. 2009], thus introducing a new class of stimulatory lipids for iNKT cells.

In order to confirm that the T cell stimulatory activity is due to formation of complexes with CD1d and not to cellular modifications induced by these piperidinones in APC, T cell activation was tested using CD1d plate-bound assays. Both L-*arabino* diols **7** and **15** were active also in this type of assay (see Figure 11.14, p.84), thus confirming that they form stimulatory complexes with CD1d. The formation of complexes with CD1d is also supported by the finding that only the L-*arabino* diols **7** and **15** possess stimulatory capacity whereas the L-*ribo* diols **8** and **16** do not (see Figure 11.13, p.83).

Importantly, iNKT cells are activated by the L-*arabino* diol **7**, but not by the L-*arabino* amino alcohol **17** and the dimethylamino alcohol **21**, with **7** differing from **17** and **21** only by the C(6)-hydroxymethyl group. A similar activity was observed for the L-*arabino* diol **15** that is active, whereas the L-*arabino* amines **18** and **22** are not. It is tempting to speculate that the primary hydroxy group is important in interacting, presumably via hydrogen bonding, with the CD1d molecule, thus determining its position within CD1d. This hydroxy group might also make binding contacts with the TCR of iNKT cells, thus stabilizing the lipid-CD1d-TCR trimolecular complex, as shown for the αGC antigen [Borg et al. 2007; Wun et al. 2008; Zajonc & Wilson 2007]. A second important finding is that only the L-*arabino* diols **7** and **15** were active, whereas the L-*ribo* diols **8** and **16** were not, suggesting that binding of the lipid to CD1d, or its positioning within CD1d, is dictated also by this structural element. Since the presence of a second lipid tail in the lipid antigen might stabilize the lipid-CD1d stimulatory complexes and facilitate T cell activation, we tested piperidinones with an additional substituent at C(3).

The L-*gluco* ceramide analogues **43**, **46a**, and **47** had a strongly stimulatory effect on iNKT cells when presented by living APC and also by plate-bound human CD1d, whereas the L-*altro* ceramide analogue **52** activated only weakly (see Figure 11.15, p.85), suggesting that also in these compounds the configuration of C(3) and C(4) is important. A similar stimulatory activity was observed in plate-bound assays (see Figure 11.16, p.86), confirming that each molecule forms stimulatory complexes with CD1d. The compounds with two lipid tails showed a similar efficacy and potency as **7** and **15**, which have only one lipid tail, indicating that addition of a short alkenyl chain at C(3) does not change this biological activity.

A final unexpected finding was that all tested compounds preferentially induce release of T_H1-like cytokines IFN-γ and TNF-α, whereas the T_H2 cytokine IL-4 is released in low amounts, as shown by the IFN-γ/IL-4 and TNF-α/IL-4 ratios with αGC, the L-*arabino* diols **7** and **15** (see Figure 11.14 (D), p.84), the L-*gluco* ceramide analogues **43**, **46a**, and **47**, and the L-*altro* ceramide analogue **52** (see Figure 11.16 (D), p.86), respectively. Thus, these piperidinones induce preferentially a T_H1 response. As the tested iNKT cells release both classes of cytokines when the αGC agonist is used as antigen, the T_H1-biased response has to be ascribed to the type of lipid-CD1d complexes formed by these piperidinones. T_H1-responses have been associated with prolonged TCR engagement, whereas T_H2-responses have been ascribed to weak interactions. It is therefore tempting to speculate that piperidinones form complexes with CD1d that make high affinity interactions with the TCR of iNKT cells and/or the complexes accumulate in detergent-resistant membrane microdomains [Im et al. 2009]. Future studies will address this point.

In conclusion, fine-tuning the structure of piperidinones may lead to the rational design of new iNKT cell activatory compounds with unique biological properties. The generation of lipid compounds preferentially inducing T_H1 responses might have applications in novel vaccination and anti-tumor therapies.

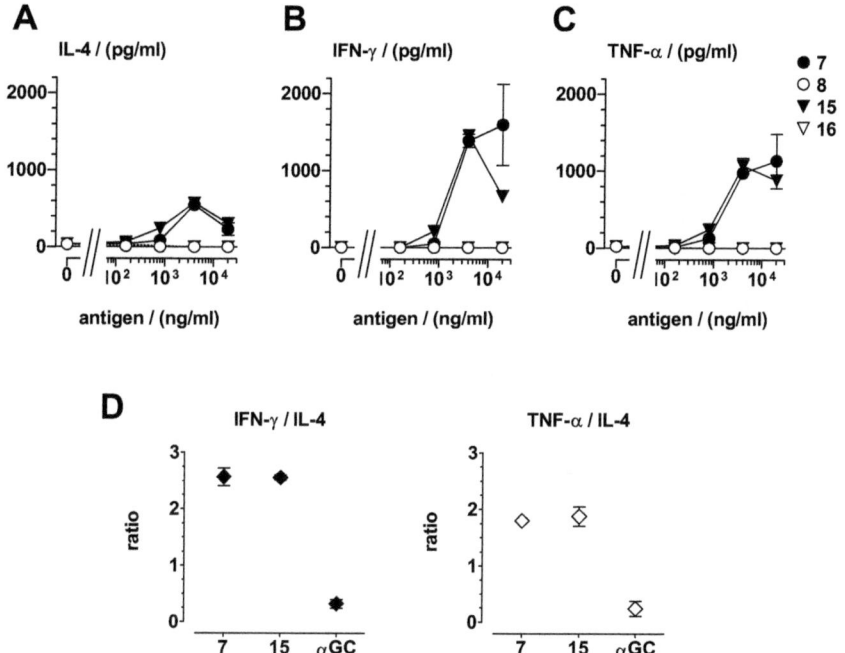

Figure 11.13
Single tail piperidinones presented by living APC activate iNKT cells. The L-*arabino* diols **7** and **15** stimulate iNKT cells when presented by THP1-hCD1d APC as measured by release of (A) IL-4, (B) IFN-γ and (C) TNF-α (closed circles = **7**, open circles = **8**, closed triangles = **15**, open triangles = **16**). (D) Diols **7** and **15** preferentially induced a T_H1 response, as visualized by the IFN-γ/IL-4 ratio of the stereotype T_H1 versus T_H2 cytokines IFN-γ and by the TNF-α/IL-4 ratio.

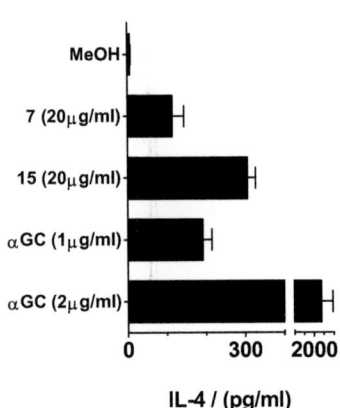

Figure 11.14
Single tail piperidinones presented by plate-bound CD1d activate iNKT cells. The L-*arabino* configured diols **7** and **15** induce cytokine production by iNKT cells when presented on recombinant soluble plate-bound human CD1d, as measured by released IL-4.

11.6. Final conclusions and outlook

Overall, the techniques introduced here will not only enable highly sensitive identification of antigen-specific iNKT cells but testing of their response towards a large variety of antigens designed as vaccines, as tumor antigens or for other applications. Broadening of these techniques to type 2 NKT cells and to other CD1-restricted T cells remains a challenge for the future.

Our finding that certain types of antigen, namely piperidinones, preferentially induce a $T_H 1$-like phenotype has to be confirmed *in vivo* as so far it has been thought that second signals determine iNKT cell function [Taniguchi et al. 2010]. iNKT cells are well know for their self-reactivity and ability to readily release large amounts of cytokines, bridging innate and acquired immunity. Homeostatic self-ligand recognition appears to prime or rather prepare, as manifested by acquisition of activation markers and accumulation of cytokine messenger RNA, iNKT cells without eliciting any effector functions. This activated phenotype of iNKT cells under steady state conditions *in vivo* ensures a rapid response to exogenous antigens. However, this might indicate that iNKT cells require additional environmental signals to mediate their functions.

As iNKT cells can functionally resemble $T_H 1$ or $T_H 2$ cells, *in vivo* manipulation may change the balance between pro- and anti-inflammatory cytokines to favor the desired outcome of the immune response depending on the disease or treatment context. In conclusion, the screening of a large number of synthetic molecules has shown that design of compounds like the piperidinones may provide optimized therapeutic reagents to induce protective or regulatory immune responses.

11.6. Final conclusions and outlook

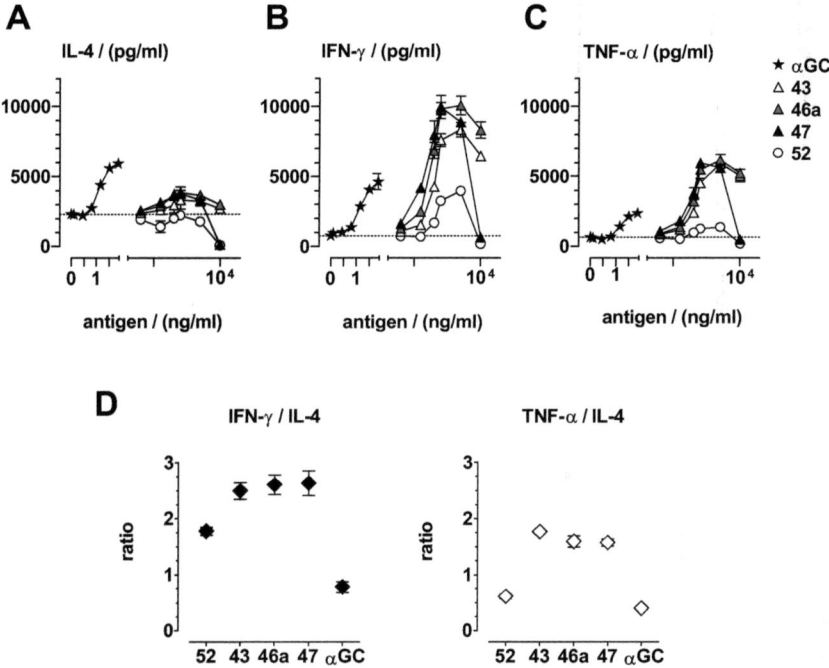

Figure 11.15
Double tail piperidinones presented by living APC activate iNKT cells. The L-*gluco* ceramide analogues **43**, **46a**, **47**, and the L-*altro* ceramide analogue **52** presented by THP1-hCD1d promote release of (A) IL-4, (B) IFN-γ and (C) TNF-α (stars = αGC, open triangles = **43**, gray triangles = **46a**, closed triangles = **47**, open circles = **52**). The dotted lines indicate background activation cytokine levels. (D) All active compounds induced more of a T_H1 response, as seen by the ratios IFN-γ and TNF-α to IL-4.

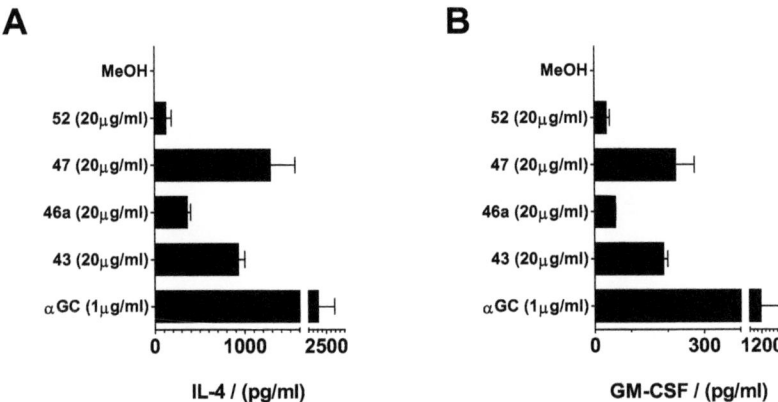

Figure 11.16
Double tail piperidinones presented by plate-bound CD1d activate iNKT cells. The L-*gluco* ceramide analogues **43**, **46a**, and **47**, and the L-*altro* configured ceramide analogue **52** are stimulatory for iNKT cells when presented on dimerized plate-bound human CD1d, as IL-4 (A) and GM-CSF (B) are released.

12. Invariant natural killer T cells link inflammation and neovascularization in human atherosclerosis

Atherosclerosis is a chronic inflammatory disease of large arteries. It is complicated by cardiovascular events which are usually provoked by plaque rupture or erosion. Inflammation participates in lesion progression and plaque rupture. Identification of the cause of inflammation and hence the leukocyte populations involved in plaque destabilization is important for effective prevention of cardiovascular events. As lipid retention supports growth of atherosclerotic plaques, lipid-specific invariant natural killer T (iNKT) cells and their antigen-presenting molecule (APM) CD1d were investigated.

In human atherosclerotic lesions, CD1d$^+$ antigen-presenting cells (APC) and iNKT cells were correlated with disease severity and activity. We aimed to elucidate potential mechanisms whereby these cells might be involved in plaque formation and/or destabilization. CD1d$^+$ APC colonized advanced plaques in symptomatic patients and were most abundant in plaques with concomitant signs of ectopic neovascularization. iNKT cells were detected in plaques and therein their frequency among the T cell compartment exceeded the one in blood. iNKT cell lines isolated from plaques promptly produced proinflammatory cytokines when stimulated by CD1d-expressing APC and alpha-galactosylceramide proving the presence of CD1d-restricted T cells. Antigen-stimulated iNKT cell-derived culture supernatants showed angiogenic activity in an endothelial cell-spheroid model of *in vitro* microvascular sprout formation and strongly activated endothelial cell migration. This functional activity was ascribed to interleukin (IL)-8 released by iNKT cells upon lipid recognition.

These findings introduce iNKT cells as novel cellular candidates to promote plaque neovascularization and destabilization in human atherosclerosis. Furthermore iNKT cell-lipid antigens with beneficial effects might be designed to be used in medical treatment or as vaccines.

Cardiovascular events which are usually precipitated by plaque rupture or erosion [Naghavi et al. 2003 a,b; Steg et al. 2007; Virmani et al. 2005] are complicating atherosclerosis (ATH), a chronic inflammatory disease of large arteries. Vulnerable plaques that are prone to rupture are characterized by spatially and temporally interconnected inflammation [Hansson 2005; Libby 2002; Redgrave et al. 2006; Ross 1999; van der Wal et al. 1994; Virmani et al. 2005], plaque hemorrhage [Kolodgie et al. 2003; Virmani et al. 2005] and abnormal apoptosis [Clarke et al. 2006; Lim et al. 2008]. Both innate and acquired immune responses can modulate atherosclerotic plaque development [Binder et al. 2002]. Macrophages and T lymphocytes infiltrating the arterial wall during atherosclerosis [van der Wal et al. 1989] are supposed to produce proinflammatory cytokines, chemokines, metalloproteinases and mesenchymal growth factors which are all potentially involved in plaque growth and rupture but might also contribute to plaque stabilization. A histopathological quantitative analysis has suggested that macrophages in the arterial wall seem to be protective in the early, but deleterious in the late stages of the disease [Fleiner et al. 2004]. T cell populations with different functional capacities have been identified within atherosclerotic lesions [van der Wal et al. 1989] and this contributes to the pathogenic complexity of the inflammatory process [Robertson & Hansson 2006].

In addition to inflammation, other mechanisms such as lipid retention [Tabas et al. 2007], neovascularization [Doyle & Caplice 2007; Moreno et al. 2006; Virmani et al. 2005] and tissue remodeling [Lijnen 2003; Newby 2006] support plaque growth. How different leukocyte populations contribute to or are affected by these additional mechanisms remains elusive. T cells recognizing protein or lipid antigens that accumulate within plaques are likely to be involved and invariant natural killer T (iNKT) cells have recently attracted attention in this respect [Major et al. 2006]. αGC, the archetype pan-iNKT cell antigen, accelerates atherosclerotic lesion formation in the mouse [Major et al. 2004; Nakai et al. 2004; Tupin et al. 2004] and CD1d-deficient as well as TCR Vα14-deficient mice, both lacking iNKT cells, are protected in this model of atherosclerosis [Aslanian et al. 2005; Rogers et al. 2008; Tupin et al. 2004]. Moreover, adoptive transfer of iNKT cells markedly increases plaque burden [VanderLaan et al. 2007]. Taken together, these animal studies provide strong evidence that iNKT cells are involved in lesion development. However, no detailed investigations have focused on iNKT cells in human atherosclerosis. Although CD1d is expressed at the protein level in human atherosclerotic lesions [Melian et al. 1999] it remains unknown whether CD1d expression correlates with lesion severity or disease activity.

Identification of leukocyte populations involved in plaque formation and destabilization is important for effective prevention and treatment of cardiovascular events. The aim of this project was to investigate CD1d-positive antigen-presenting cells (APC) and iNKT cells in human atherosclerotic lesions, correlate them with disease severity and activity, and elucidate potential mechanisms by which these cells might be involved in plaque formation, progression and destabilization. Lesional CD1d$^+$ APC were present in large quantities in advanced plaques and were most abundant in plaques with accompanying ectopic neovascularization. iNKT cells were detected inside plaque tissue and were enriched in frequency among CD3$^+$ T cells when compared to blood. For the first time phenotypical iNKT cell lines were isolated from plaques. The isolated iNKT cell lines were able to promptly and efficiently produce large amounts of proinflammatory cytokines upon stimulation by CD1d-expressing APC presenting αGC, thus proving the presence of CD1d-restricted T cells. Furthermore, their antigen-specific activation threshold is reduced by at least one log compared to iNKT cells from blood of healthy donors, meaning the antigen potency is increased tremendously. Cell culture supernatants derived from antigen-stimulated iNKT cells showed angiogenic activity in an endothelial cell-spheroid model of *in vitro* microvascular sprout formation and strongly induced endothelial cell migration. This functional influence on endothelial cell activity was imputed to interleukin (IL)-8 released by iNKT cells upon lipid recognition.

12.1. CD1d-expressing cells in atherosclerotic lesions are a sign of arterial vulnerability

Macrophages and CD1d$^+$ cells were quantified in arterial tissue (see Figure 12.1, p.91) obtained from asymptomatic patients (n = 27) and patients with symptomatic disease (n = 22) using human arterial tissue microarrays [Fleiner et al. 2004]. With this technique we investigated a total number of 196 arterial sectors. Each sector was histologically scored for plaque type according to the American Heart Association (AHA) consensus report [Stary et al. 1992] and the presence of CD1d$^+$ cells and CD68$^+$ macrophages as well as von Willebrand factor-positive microvessels was determined. Cell numbers were determined per intimal area to enable appropriate comparison between sectors. This approach allowed us to correlate histomorphological findings with disease activity and lesion severity. The clinical characteristics of the 49 patients are shown in a table (see Table 9.2, p.53). Both CD68$^+$ macrophages and CD1d$^+$ cells were more commonly found in advanced lesions than in early plaque stages (see Figure 12.1 (A), p.91). We found CD68$^+$ macrophages (P=0.029) and CD1d$^+$ cells (P<0.001) more frequently in plaques from symptomatic patients with previous cardiovascular events (see Figure 12.1 (B), p.91). In patients with symptomatic, clinically active disease, atherosclerotic plaques with ectopic neovascularization (neovessels) [Fleiner et al. 2004] had on average the highest numbers (P<0.001 and P=0.002 respectively) of macrophages and CD1d$^+$ cells (see Figure 12.1 (B), p.91). Among the patients with symptomatic ATH, 16 died of a incidental cardiovascular event. The number of CD1d$^+$ cells inside the intimae of these 16 patients (4.24 (0.72-11.55) cells/mm2) was comparable to that of the 6 patients with stable cardiovascular disease (2.25 (0.74-31.91) cells/mm2). This suggests that CD1d$^+$ cell accumulation represents an irreversible, rather than a transient process of plaque destabilization. It is remarkable that CD1d$^+$ cells are virtually absent from early lesions and are uncommon in asymptomatic patients. For the tissue microarray analysis, arterial rings were harvested on average 24 h after death. We tested whether the number of detectable CD1d$^+$ cells would diminish with progressive time after death and found no correlation between these two parameters (data not shown). Since the presence of CD1d$^+$ APC in the arterial intima positively correlates with lesion severity, disease activity and neovascularization, these cells may participate critically in plaque growth and destabilization.

12.2. iNKT cells are found in atherosclerotic lesions

The presence of CD1d$^+$ APC in advanced, unstable atherosclerotic lesions prompted a search for iNKT cells. Due to the predicted scarcity of these cells, several different approaches were applied to demonstrate their presence in atherosclerotic plaques. Lesional arterial intima from 5 patients with ATH was examined by confocal microscopy. We demonstrated the presence of CD3$^+$/Vα24$^+$ and CD3$^+$/Vβ11$^+$ cells, which represented up to 3% of total infiltrating CD3$^+$ T cells in all lesions analyzed [Kyriakakis et al. 2009]. Since these findings only suggest but do not prove the presence of iNKT cells we next prepared cell suspensions from thrombendarterectomy specimens and performed co-staining with anti-Vα24 and anti-Vβ11 mAb [Kyriakakis et al. 2009]. The identification of Vα24/Vβ11 double-positive (DP) cells provided evidence that iNKT cells are present in the diseased arterial wall.

We additionally performed fluorescence and confocal microscopy of lesional tissue from 20 patients with anti-CD1d and anti-TCR Vα24-Jα18 mAb, which recognizes the iNKT-specific pairing of the TCR α-chain variable and joining region namely Vα24 to Jα18 [Exley et al. 2008; Montoya et al. 2007]. Representative micrographs unequivocally demonstrate the presence of iNKT cells in atherosclerotic lesions (see Figure 12.2, p.93). In tissues incubated with an iNKT cell clone to allow their invasion, TCR Vα24-Jα18 staining was mostly seen within lesional

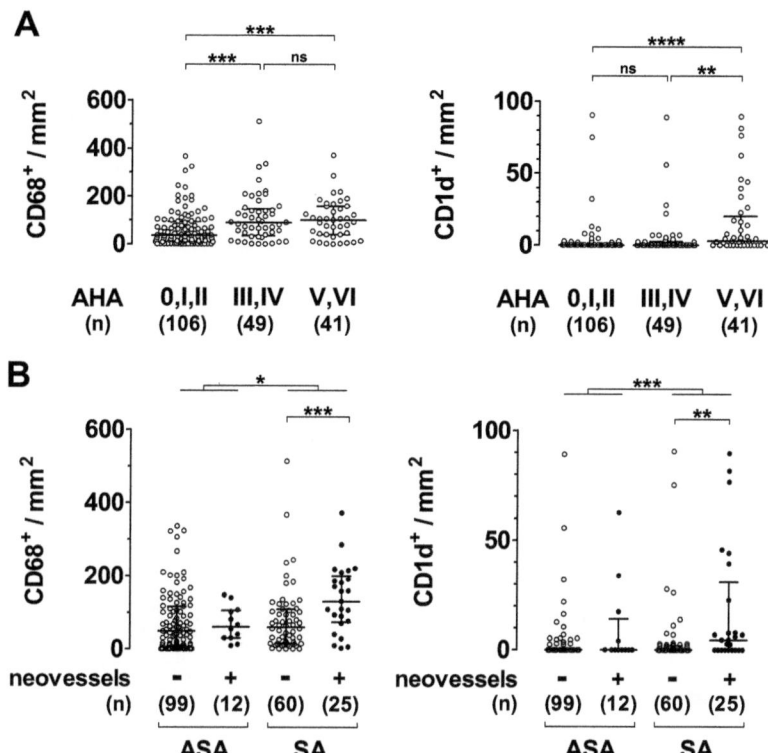

Figure 12.1
APC in atherosclerotic lesions. APC in atherosclerotic lesions. The number of CD68$^+$ macrophages and CD1d$^+$ cells per intima area were determined with arterial tissue microarrays. For each arterial sector, the intima area was morphometrically measured and the CD1d$^+$ and CD68$^+$ cells in the intima counted (expressed as cells/mm2). Each data point represents a single sector. In addition, median values are shown and error bars represent the interquartile range. (A) Macrophage and CD1d$^+$ cell counts in arterial sectors affected by atherosclerotic lesions of increasing severity according to the AHA classification. (B) Quantitative analysis of macrophage and CD1d$^+$ cell counts in arterial sectors according to disease activity (i.e. whether patients suffered cardiovascular events during their lifetime or not and said to have asymptomatic (ASA) or symptomatic ATH (SA)). Neovessels were detected as von Willebrand factor-positive microvessels in the arterial intima and plaques were scored positive (closed circles) or negative (open circles) with respect to this anatomical sign. * $P<0.5$, ** $P<0.01$, *** $P<0.001$, **** $P<0.0001$, ns = not significant. Experiments were performed in part by Jan Andert.

vasa vasora/neovessels or adjacent tissue but not deep within the tissue as observed for tissue sections of patients. Furthermore, there was evidence of iNKT TCR and CD1d polarization towards each other and even colocalization of the iNKT TCR with CD1d (see Figure 12.3, p.94). These findings indicate an ongoing activation of iNKT cells on CD1d$^+$ APC deep inside the atherosclerotic tissue.

To confirm and formally prove that iNKT cells reside within atherosclerotic lesions we next isolated and expanded iNKT cells from thrombendarterectomy specimens containing plaques obtained from symptomatic patients and performed phenotypic and functional studies. We established 6 bulk T cell lines by stimulation with αGC and CD1d$^+$ APC. Flow cytometry analysis using 5 color staining showed that 60-90% of CD3$^+$ cells were co-expressing TCR Vα24 and Vβ11 chains (see Figure 12.4 (A), p.95). In all lines, Vα24$^+$/Vβ11$^+$ cells were also stained with αGC-loaded soluble human CD1d dimers and five out of the six iNKT cell isolates expressed the CD4 molecule, as shown by one representative line (see Figure 12.4 (A), p.95), and one was double-negative (DN) for CD4 and CD8. When plaque-derived iNKT cells were stimulated with αGC and CD1d$^+$ APC [Brossay et al. 1998], they produced large amounts of IL-4, TNF-α, IFN-γ and GM-CSF (see Figure 12.4 (B), p.95), resembling iNKT cells from peripheral blood. To compare responsiveness of plaque-derived iNKT cell lines to circulating iNKT cells, we calculated the half maximal effective concentration (EC$_{50}$) values of αGC of the established lines and a set of iNKT cell clones from peripheral blood of different normal donors when releasing IL-4, IFN-γ (see Figure 12.5 (C), p.96), TNF-α and GM-CSF (data not shown). For iNKT cell lines from plaques the EC$_{50}$ values of αGC were at least 10-fold lower as compared to peripheral blood-derived iNKT cell clones, pointing to a high responsiveness of plaque-derived iNKT cells, at least to this lipid antigen. Taken together, these results prove that the iNKT cells present within atherosclerotic lesions have phenotypic and functional features of bona fide iNKT cells [Godfrey & Kronenberg 2004 a] and react to αGC with unusual high efficiency maybe due to an antigen selection mechanism inside plaques. The mechanism by which the antigen-specific activation threshold of iNKT cells is lowered in or by plaques remains unknown and needs further investigation.

12.3. Reduction of circulating iNKT cells in ATH patients

Next we investigated whether a similar expansion of iNKT cells was detectable in the circulating blood of vulnerable patients with symptomatic ATH. Three groups of donors, namely patients with symptomatic ATH, age-matched control patients free of cardiovascular events in the past (i.e. asymptomatic) and young healthy control individuals free of any sign of atherosclerotic disease were compared. Single living iNKT cells were detected in peripheral blood mononuclear cells (PBMC) [Mutschelknauss et al. 2007] with four color immunofluorescence analysis using anti-CD3, anti-TCR Vα24, anti-TCR Vβ11 mAb and αGC-loaded CD1d dimers. Unexpectedly, we detected a profound reduction of circulating iNKT cells in symptomatic patients (P=0.001) compared with the two control groups (see Figure 12.6, p.96). A reduction was also observed in elderly patients free of cardiovascular events (P<0.01), possibly reflecting an age-related effect on this lymphocyte subset [Jing et al. 2007]. When taken together with the colonization of atherosclerotic lesions by CD1d$^+$ cells (see Figure 12.1 (B), p.91), our observations are in accordance with the concept that iNKT cells undergo tissue redistribution and perhaps home preferentially to atherosclerotic lesions in patients with clinically active disease.

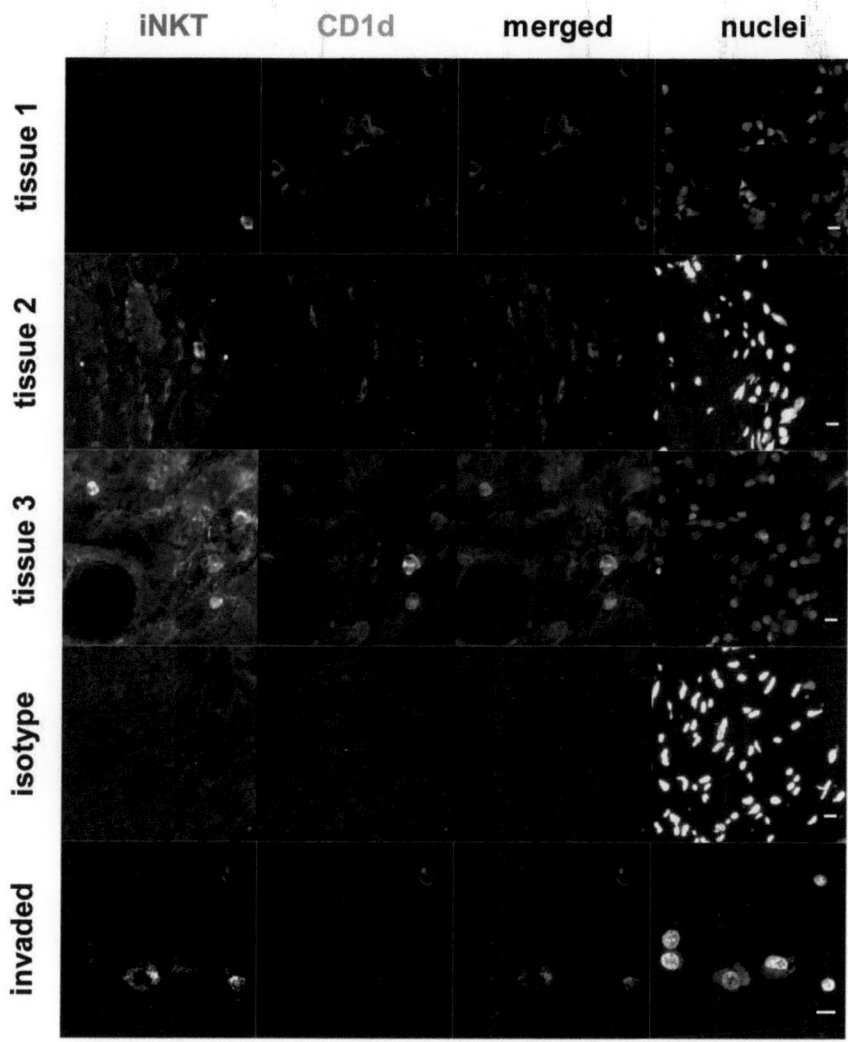

Figure 12.2
Specific identification of iNKT cells in ATH lesions. Columns (left to right) show the staining of iNKT cells with anti-TCR Vα24-Jα18 (6B11), of CD1d$^+$ cells (51.1), iNKT in red merged to CD1d$^+$ cells in green (merged), and of nuclei (Hoechst). The first three rows from the top show staining of tissue from three different patients. The fourth row gives a representative isotype staining of tissue from one of the three patients. The fifth row depicts a staining of a section from arterial tissue invaded by an iNKT cell clone. Scale bars represent 10 µm.

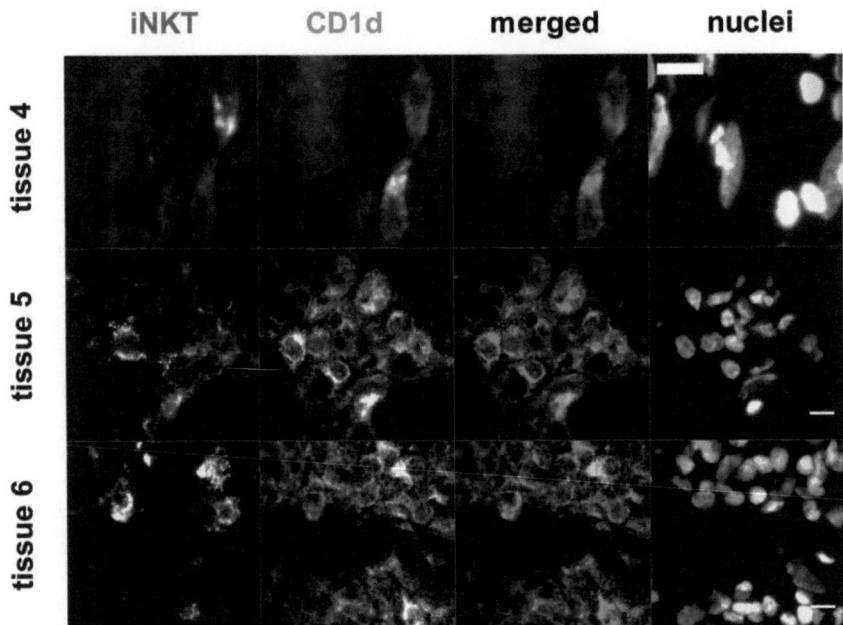

Figure 12.3
Polarization of the TCR on iNKT cells towards CD1d on APC in ATH lesions. Columns (left to right) show the staining of iNKT cells with anti-TCR Vα24-Jα18 (6B11), of CD1d$^+$ cells (51.1), iNKT in red merged to CD1d$^+$ cells in green (merged), and of nuclei (Hoechst). The rows show staining of three representative tissue sections of two individual patients demonstrating polarization and/or colocalization of the TCR on iNKT cells towards CD1d on APC and vice versa. Scale bars represent 10 µm.

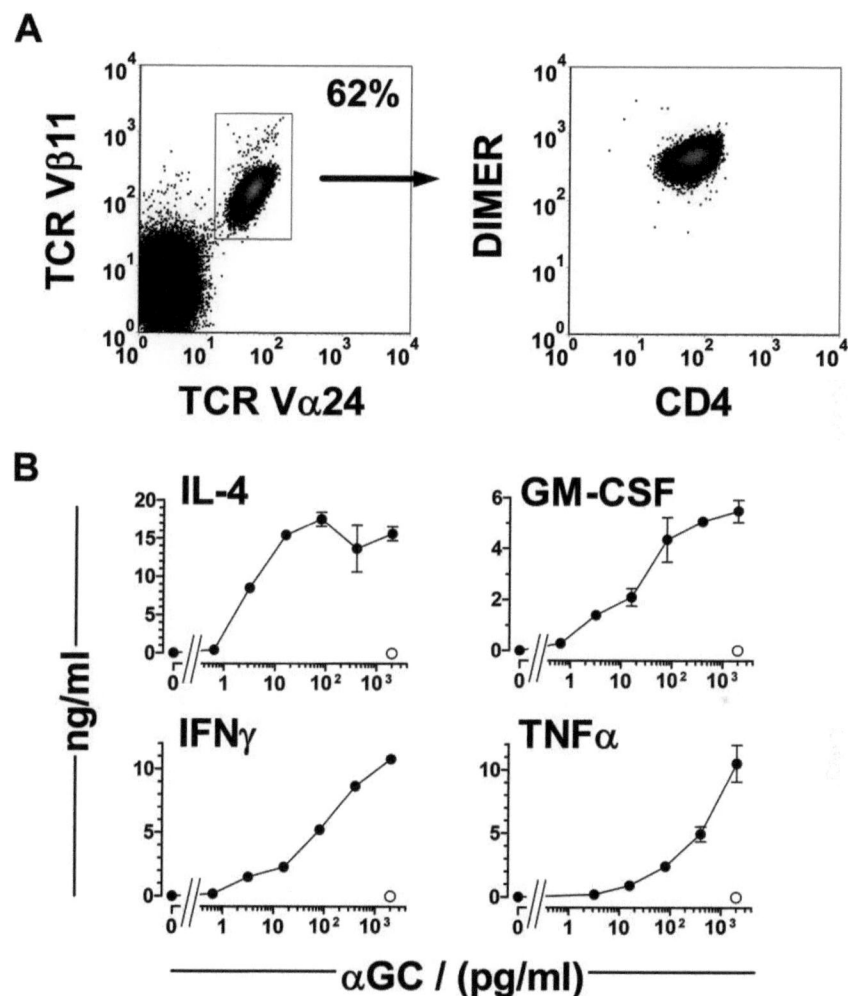

Figure 12.4
iNKT cells from atherosclerotic plaque tissue. (A) A representative (of 6 total) bulk T cell line isolated from plaques and expanded after stimulation with αGC. T cells were stained with anti-CD3, -CD4, -CD8 (not shown), -TCR Vα24 and -TCR Vβ11 mAb and with αGC-loaded CD1d dimers (DIMER). Density plots gated on CD3$^+$ (left) and on CD3/TCR Vα24/TCR Vβ11 triple-positive cells (right) are shown. (B) Lymphokine release from one representative plaque-derived iNKT cell line upon *in vitro* stimulation with αGC in the presence (closed circles) or absence (open circles) of CD1d$^+$ APC. Results are expressed as mean ng/ml of triplicates ± SD. Similar results were obtained with the other 5 cell lines in at least two independent experiments.

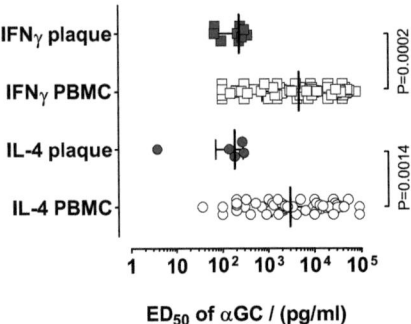

Figure 12.5

Half maximal effective concentration of alpha-galactosylceramide with iNKT cells from atherosclerotic plaque tissue. Potency of αGC on iNKT cell clones established from PBMC (open symbols) or with iNKT cell lines isolated from plaque tissue (closed symbols). iNKT cells were added after αGC titration on APC, cytokine release assessed by ELISA and EC_{50} values calculated. Each symbol represents the EC_{50} value of one titration experiment. Median and interquartile range are given per group. For all cytokines measured the potency of αGC with plaque-derived iNKT cells is at least 1 log lower than with iNKT cells from PBMC (P-values are given beside the brackets).

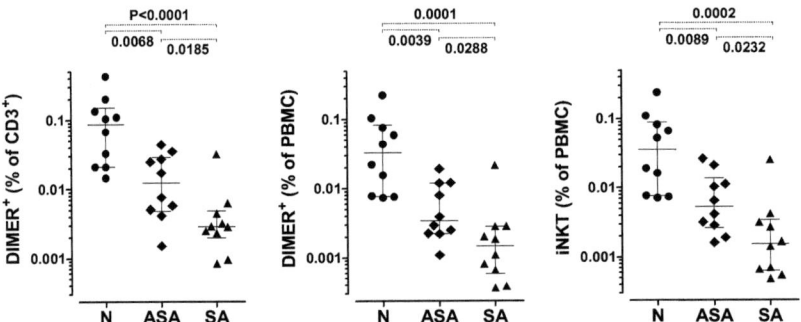

Figure 12.6

Circulating iNKT cells are reduced in ATH patients. Distribution of iNKT cells in PBMC from healthy young donors (N) and from age-matched patients with asymptomatic ATH (ASA) or patients with symptomatic ATH (SA). iNKT cells were detected by FACS with αGC-loaded CD1d dimers ($DIMER^+$) or with anti-TCR Vα24 and anti-TCR Vβ11 mAb (iNKT). Data are shown as percentage after gating on T cells by means of CD3 or as percentage of total PBMC. P-values are given above brackets.

12.4. Characterization of proatherosclerotic activity of iNKT cells

The presence of CD1d$^+$ cells and iNKT cells within advanced atherosclerotic lesions, particularly in patients with symptomatic disease, prompted us to investigate whether this T lymphocyte population has a role in key processes of plaque formation and destabilization. Given the fact that plaque-derived iNKT cell lines were prone to cytokine release (see Figure 12.6 (C), p.96), we investigated the spectrum of cytokines and chemokines produced by antigen-stimulated iNKT cells in a BioPlex assay and found the presence of a number of proinflammatory and potential angiogenic modulators [Kyriakakis et al. 2009]. Since neovascularized arterial sectors had the highest numbers of CD1d$^+$ cells, subsequent investigations focused on effects of iNKT cell activation on angiogenic behavior of endothelial cells (EC).

We examined angiogenic potential of conditioned medium (CM) derived from iNKT cell cultures stimulated with (CM+) or without (CM−) αGC using the EC-spheroid model of *in vitro* microvascular sprout formation as a global functional test for angiogenesis. Visualization of spheroids composed of human microvascular EC indicated that CM+ induced greater sprout outgrowth than CM− (see Figure 12.7 (A), p.98). Morphometric analysis showed a significant increase in both the number (see Figure 12.7 (B), p.98) and length (see Figure 12.7 (C), p.98) of sprouts. CM collected from cultures containing αGC but lacking either CD1d$^+$ APC or iNKT cells, or both failed to enhance sprout outgrowth [Kyriakakis et al. 2009]. Taken together, these data confirm that antigen-stimulated iNKT cells can promote angiogenesis *in vitro*.

12.5. Soluble factors released by iNKT cells promote EC migration

Angiogenesis is a complex process and both proliferation and migration of EC contribute to this phenomenon [Adams & Alitalo 2007; Doyle & Caplice 2007]. To identify which of these activities is modulated in response to iNKT cell activation, we compared effects of CM on proliferation and migration of EC in monolayer cultures. CM+ derived from different iNKT cells did not activate EC proliferation [Kyriakakis et al. 2009] but did induce cell migration. Two different methods were used to evaluate migration. Confluent EC monolayers were scrape-wounded and EC migration into the wound was recorded over a 12 h period by time lapse videomicroscopy. This wound healing assay showed more rapid migration into the wound area for EC cultured in the presence of CM+ (see Figure 12.8, p.99). Representative videos showing EC motility in the presence of CM− (FigS4video1-CM−.avi) and CM+ (FigS4video2-CM+.avi) are given in the Supplemental Material Appendix [Kyriakakis et al. 2009]. The second assay quantified transmigration of EC in a Boyden chamber and also demonstrated enhanced migration of EC toward CM+ (see Figure 12.8 (C), p.99). These data suggest a chemokine-dependent effect on EC angiogenic behavior.

12.6. IL-8 is produced by iNKT cells and induces EC migration

IL-8, a pleiotropic chemokine with known angiogenic activity *in vitro* and *in vivo* [Li et al. 2003; Simonini et al. 2000], was amongst numerous factors released by activated iNKT cells [Kyriakakis et al. 2009]. iNKT bulk lines isolated from plaques (see Figure 12.9 (A), p.101) and iNKT cell clones [Kyriakakis et al. 2009] showed strong intracellular staining for IL-8 when stimulated with αGC proving that they readily produce this chemokine. To determine the contribution of iNKT cell-released IL-8 to the angiogenic potential of CM+, wound-healing assays were conducted in the presence of anti-IL-8 blocking antibodies or using CM which had been subjected to immunodepletion of IL-8 prior to assay. Both treatments completely abrogated the

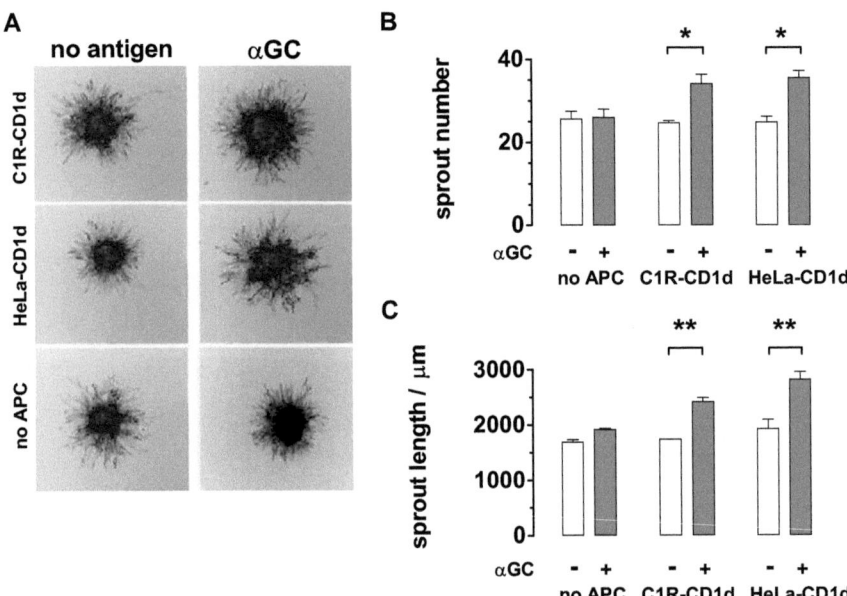

Figure 12.7
Antigen activation of iNKT cells increases sprout outgrowth from EC spheroids. Conditioned medium derived from iNKT cell cultures stimulated with (CM+) or without (CM−) αGC was examined using the EC-spheroid model of *in vitro* angiogenesis. (A) Representative images of whole spheroids 24 h after exposure to CM. Morphometric analysis of spheroids for total sprout number (B) and total sprout length (C) shows significantly greater angiogenic potential for CM+ (filled bars) compared with CM− (open bars). Bars undermarked 'no APC' indicate the response to CM from iNKT cells cultured alone. Data are mean ± SD from six independent experiments (* $P<0.05$, ** $P<0.01$). CM+ from a second iNKT cell clone elicited similar proangiogenic effects (data not shown). Experiments were performed in part by Emmanouil Kyriakakis

12.6. IL-8 is produced by iNKT cells and induces EC migration

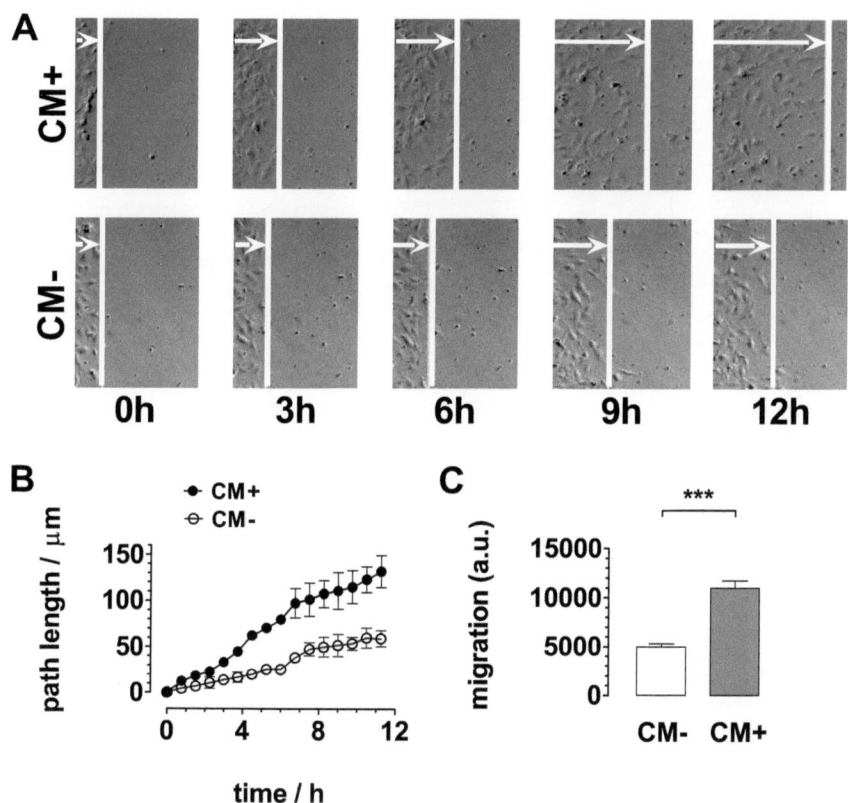

Figure 12.8
Antigen activation of iNKT cells promotes EC migration. Confluent monolayers of EC were scrape-wounded and the subsequent rate of wound closure monitored over a time period of 12 h by time lapse videomicroscopy. Acquired images were processed and analyzed using cell^R software. (A) Representative images of EC migration acquired at 0, 3, 6, 9 and 12 h after culture of wounded EC monolayers in the presence of CM+ or CM−. White lines indicate the location of the wound front and arrows indicate migration path length. (B) Quantitative analysis of the rate of EC migration from the initial wound front into the wound area (path length versus time). The data are representative of at least 30 independent experiments, each one performed in duplicate and values are given as averaged path length measurements ± SD from triple fixed observation fields/well. Motility was similarly enhanced by CM+ from different iNKT clones (data not shown). (C) EC transmigration toward CM+ and CM− in Boyden chamber chemotaxis assay. CM was present in the lower wells of the chamber. EC in serum-free culture medium were seeded into the upper wells and transmigration quantified after a 6 h incubation. Data are reported as mean ± SD (in arbitrary units) from three separate experiments each performed in duplicate. *** $P<0.001$. Experiments were performed in part by Emmanouil Kyriakakis

enhanced EC migration (see Figure 12.9 (B), p.101). Neither the inclusion of anti-IL-8 antibodies nor IL-8-depletion affected the basal EC migration, thus excluding non-specific inhibitory effects of the antibodies. Therefore, the enhanced migration response of EC to CM+ is dependent upon IL-8 released by activated iNKT cells.

12.7. Discussion and Outlook

Our study investigated iNKT cells in human ATH. We found that cells expressing CD1d, the lipid-antigen-presenting molecule (APM) for iNKT cells, are more abundant in advanced atherosclerotic plaques and in lesions from patients with active, symptomatic disease. Amongst the patient group with symptomatic disease, vascularized plaques had the highest number of CD1d$^+$ cells. We identified the presence of iNKT cells in atherosclerotic lesions and characterized their function after isolation from plaques. iNKT cells from plaques show a high reactivity to the αGC antigen and may promote neovascularization via an IL-8-dependent mechanism. Our study indicates that iNKT cells may contribute to the predisposition of atherosclerotic plaques to rupture.

In order to perform a detailed and quantitative immunohistochemical analysis of inflammatory cells in atherosclerotic plaques, we took advantage of tissue microarrays which allowed evaluation of 196 samples from 49 patients. This technical approach facilitated the correlation of CD68$^+$ macrophages and CD1d$^+$ cells with lesion severity (i.e. plaque stage) and with plaque neovascularization. Whereas CD68$^+$ macrophages are commonly found in the human arterial wall regardless of the plaque stage [Fleiner et al. 2004], CD1d$^+$ cells were virtually absent from the normal arterial intima or in early plaque stages. Expression of CD1d in human atherosclerotic plaques has been reported in two studies [Bobryshev & Lord 2005; Melian et al. 1999]. However, only a small number of samples was analyzed and a correlation was not made with the clinical stage and neither with disease activity nor with histological hallmarks. Thus the question of whether CD1d expression correlates with lesion grade was left open. In our study a large number of CD1d$^+$ cells was observed in advanced lesions (AHA type >IV) and particularly in lesions with signs of neovascularization, thus demonstrating a close correlation with advanced disease. We found that in asymptomatic patients, CD1d expression was present in only a minority (<20%) of arterial sectors. In contrast, in patients with symptomatic ATH CD1d$^+$ cells were found more frequently and in up to 65% of the sectors with concomitant signs of neovascularization. The preferential localization of CD1d$^+$ cells in areas with neovascularization could be explained by their efficient recruitment into vascularized plaques [Moulton et al. 2003] or, alternatively, by their capacity to promote plaque neovascularization. Our data are in accordance with the concept that CD1d may present lipid antigens locally to specific T cells, including iNKT cells, which in turn may release angiogenic factors and contribute to neovascularization.

In the diseased arterial wall we found T cells expressing Vα24 or Vβ11 TCR chains, which are used by iNKT cells. The identity of iNKT cells was confirmed by detection of TCR Vα24-Jα18, the iNKT cell-specific TCR α-chain variable and joining region pairing, in tissue sections, by detection of Vα24 and Vβ11 co-expressing T cells freshly isolated from lesions, and by expansion of iNKT cell lines from plaques. The intraplaque infiltration of iNKT cells is accompanied by their significant reduction in circulating blood in symptomatic patients. Although we could not directly compare the iNKT cell numbers in plaques and circulating blood from the same individuals, our results suggest that in plaques there is a 10-100 fold enrichment of iNKT cells relative to T cells. This unusual accumulation might be caused by preferential homing to plaques and retention following local activation and/or local proliferation upon antigen recognition.

The presence of iNKT cells in lesions has been inferred in mouse ATH models by molecular investigations and not cellular isolation. In a pioneering study on ApoE-deficient mice under high cholesterol diet, the presence of iNKT cells was suggested by RT-PCR studies [Paulsson

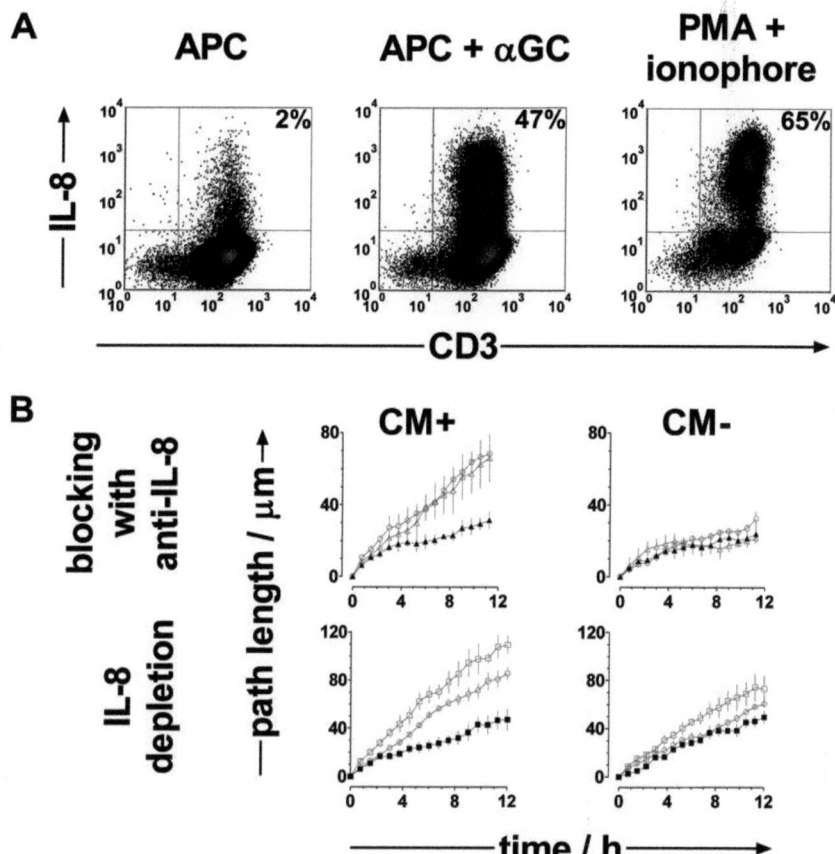

Figure 12.9

Antigen activated iNKT cells produce IL-8 which promotes EC migration. (A) Intracellular IL-8 analyzed by FACS in plaque-derived lines either resting (APC alone) or activated with APC + αGC or with PMA + ionophore. Cells were stained intracellularly with anti-IL-8 and anti-CD3 mAb. Percentages of CD3$^+$ cells producing IL-8 are indicated. (B) Effects of inclusion of anti-IL-8 blocking mAb (upper panels) and of IL-8 immunodepletion (lower panels) of CM on EC migration examined by wound assay. Upper graphs: assays were performed in the absence (open circles) and presence of neutralizing anti-IL-8 mAb or isotype control IgG (closed or open triangles, respectively). Lower graphs: assays were performed with untreated CM (open circles), or CM subjected to immunodepletion protocols using neutralizing anti-IL-8 mAb or isotype control IgG (closed or open squares, respectively). Data shown are the average path length measurements ± SD from triple fixed observation fields/well of duplicate samples. Similar results were obtained in three independent experiments using CM from different iNKT cell clones. Experiments were performed in part by Emmanouil Kyriakakis

et al. 2000]. This observation was more recently confirmed using the same technique [Major et al. 2004; Nakai et al. 2004]. Nevertheless, to date iNKT cells have neither been isolated from plaques nor functionally characterized. In one study, CD3$^+$/CD161$^+$ cells were histologically detected in carotid specimens and appeared with a frequency of 0.3-2% among plaque infiltrating T cells [Bobryshev & Lord 2005]. Yet, CD161 is expressed by a variety of T lymphocytes including iNKT cells and therefore is not a specific marker for this latter population. In a second study, CD3$^+$ cells were expanded from aortic aneurysms with polyclonal activation [Chan et al. 2005]. However, all expanded cells expressed the CD161 marker, suggesting an abnormal proliferation of this cell type in vitro. Since the expression of semi-invariant TCR Vα24/Vβ11 was not further investigated the presence of iNKT cells was not confirmed. We isolated plaque-infiltrating iNKT cells which all were Vα24$^+$ and Vβ11$^+$. The fact that they also bound αGC-loaded CD1d dimers is clear evidence that they are classical (type I) iNKT cells. These cells were also efficiently activated by αGC presented by CD1d-expressing APC and released large amounts of cytokines, including IL-4, IFN-γ, GM-CSF and TNF-α. Thus, in atherosclerotic plaques there is accumulation of cells expressing phenotypic and functional features of bona fide iNKT cells.

Importantly, the iNKT cell lines isolated from plaques all showed an extremely low threshold of activation when stimulated with αGC. The same low threshold was instead found only in a minor fraction of iNKT cell clones isolated from peripheral blood. These findings might suggest that in plaques there is a preferential accumulation of iNKT cell populations expressing high affinity TCR. Another possibility is that plaque-derived iNKT cells lack expression of natural killer (NK) inhibitory receptors and therefore they are activated by very low doses of antigen. Recently, it has been reported that in a model of atherosclerosis in ApoE-deficient mice iNKT cells with low expression of the inhibitory Ly49 receptors show proatherogenic activity, which is more pronounced than that of Ly49-positive iNKT cells [To et al. 2009]. Future studies will compare the expression of inhibitory receptors on iNKT cells isolated from plaques and peripheral blood of the same donors.

How iNKT cell activation has a proatherogenic effect remains an open issue. One potential mechanism relates to inflammation. Human atherosclerosis is also characterized by progressive inflammation; since human iNKT cells isolated from plaques do release proinflammatory cytokines, potentially in response to very low amounts of antigen, their chronic in situ activation by lipid antigens might lead to lesion progression. This hypothesis is in line with many studies conducted in mice. Injection of the potent iNKT cell agonist αGC increases the size and number of plaques in a mouse ATH model [Tupin et al. 2004]. This experimental manipulation of iNKT cell activation elicits a massive release of type 1 helper T cell (T$_H$1) and type 2 helper T cell (T$_H$2) cytokines and an elevation in plasma levels of IL-6 and monocyte chemoattractant protein 1, which have been proposed to enhance local inflammation [Major et al. 2006; Tupin et al. 2004].

A second pathogenic mechanism may relate to neovascularization. The significant association of CD1d$^+$ cells with neovascularization in plaques suggested that iNKT cells may be directly involved in angiogenic processes. Our findings revealed that iNKT cell activation by antigen has proangiogenic effects as shown by enhanced microvascular sprout formation in an in vitro assay of angiogenesis. This effect was not associated with EC proliferation but with EC migration as demonstrated by enhanced EC motility in both wound healing and transmigration Boyden chamber assays. Amongst the multiple cytokines that were produced by activated iNKT cells, IL-8 was the most promising candidate to further investigate in this context. iNKT cells isolated from plaques produce IL-8 as shown by intracellular staining. IL-8 was also detected previously in the supernatant from lipid-stimulated blood-derived iNKT cells [Chang et al. 2007]. We found that enhanced EC migration was dependent on IL-8 produced by iNKT cells. The experiments with IL-8-blockade or IL-8 immunodepletion showing abrogation of the response support this conclusion but do not rule out that co-stimulating agonists of IL-8 signaling are present in the culture media. Furthermore, these *ex vivo* studies do not prove that iNKT cell-derived IL-8 is directly involved in plaque neovascularization and this hypothesis should be tested in appropriate animal models. The participation of IL-8 in atherosclerotic lesion progression is suggested by several studies [Reape & Groot 1999]. IL-8 has been detected in atheromatous tissue [Apostolopoulos et al. 1996; Simonini et al. 2000; Wang et al. 1996] and can be induced in monocytes by oxidized LDL and cholesterol [Terkeltaub et al. 1994; Wang et al. 1996]. Functionally, IL-8 contributes to intimal macrophage accumulation [Boisvert et al. 1998], to endothelial adhesiveness for monocytes [Gerszten et al. 1999], has mitogenic and chemoattractant effects on smooth muscle cells [Yue et al. 1994] and may also facilitate plaque recruitment of $CD8^+$ effector T cells with high cytotoxic potential [Hess et al. 2004]. IL-8 has been proposed as an important mediator of angiogenesis in cardiovascular lesions contributing to plaque growth [Simonini et al. 2000]. It is tempting to speculate that iNKT cells, when locally activated by lipid antigens, exert both promigratory and proinflammatory functions which may become important for plaque neovascularization and destabilization. These functions might be shared with resident monocytes and other T cells recognizing specific antigens in plaques. At present, the endogenous lipid antigens stimulating iNKT cells are unidentified. If the lipid antigens would be known, their detection within atherosclerosis lesions would support a cause-effect relationship in man.

In conclusion, our studies have revealed $CD1d^+$ cells in advanced, vascularized atherosclerotic lesions from patients with active disease. We have identified the presence of iNKT cells within plaques. We have isolated plaque iNKT cells and demonstrated their high sensitivity to antigen stimulation and proinflammatory and proangiogenic potential *in vitro*. By these mechanisms, iNKT cells could participate in plaque growth and destabilization. The hypothesis recently emerged that atherosclerosis is an autoimmune disease and immunotherapeutic approaches were discussed [Hansson & Libby 2006]. In the context of a potential T memory stem cell population induced by the chronic progression of ATH, our observations might introduce iNKT cells, and $CD1d^+$ APC, as potential candidate targets for immuno-preventative and -therapeutic interventions in the future. Furthermore, identification of the lipid antigens involved in iNKT cell activation would enable novel types of immunomodulatory treatments and thereafter a direct demonstration of a cause-effect relationship in man by vaccination.

13. Rafting and docking with CD1a

In professional antigen-presenting cells (APC) like dendritic cells (DC), CD1a is mainly expressed at the cell surface and its organization within membranes is poorly understood. Our studies report that major histocompatibility complex (MHC) class II, invariant chain (Ii) and CD9 are coimmunoprecipitated with CD1a in immature DC and that CD1a/invariant chain colocalization is dependent on raft integrity. Expression of CD1a in priorly CD1a-negative cells leads to increased invariant chain trafficking to the cell surface. Silencing of invariant chain in DC induces CD1a accumulation at the plasma membrane while the overall CD1a expression remains similar. Localization of CD1a in lipid rafts is functionally relevant as raft disruption inhibits CD1a-restricted antigen presentation. These findings identify invariant chain and lipid rafts as key regulators of CD1a organization on the surface of immature DC and of its immunological function as antigen-presenting molecule (APM).

Having elucidated the surface appearance, organization and internalization of CD1a, its intracellular routes and molecular trafficking mechanisms are still elusive. A major step in understanding differences in the nature of antigen presentation was the realization that MHC class I molecules sample peptides transported to the endoplasmic reticulum (ER) from the cytosol while class II molecules sample peptides generated within lysosomes. Among the CD1 isoforms, CD1a is unique because it lacks any sorting motifs (e.g. a tyrosine-based cytoplasmic sorting motif, responsible for internalization by the clathrin-mediated pathway) and solely localizes to the early endocytic and recycling compartments.

Our investigations of CD1a trafficking and the role of its short cytoplasmic tail revealed that CD1a can be internalized by a clathrin- and dynamin-independent pathway and that it follows a Rab22a- and adenosine diphosphate ribosylation factor 6-dependent recycling pathway, similar to other clathrin-independent cargo. Post-translational S-acylation of the CD1a cytoplasmic tail was found to occur but neither determines the rate of internalization nor recycling nor its localization to detergent-resistant membrane microdomains. These findings place CD1a closer to MHC class I in its trafficking and potential antigen-loading compartments among CD1 isoforms.

As each of the five members of the CD1 family (CD1a-e) localizes to a distinct subcompartment of endosomes, it has been widely assumed that the distinct trafficking of CD1 isoforms must have evolved to enable them to survey lipid antigens that follow different routes. Predictably, CD1a might have evolved to cope with lipids that are contained in the early endocytic and recycling compartments. Strikingly, the glycolipid antigen sulfatide also localizes almost exclusively to these compartments. Swapping the cytoplasmic tail of CD1a for the one of CD1b and hence targeting it to the late endosomal and lysosomal compartments decreased its capacity to present antigen and shortened the half-life of stimulatory complexes. Thus, the intracellular trafficking route of CD1a is essential for efficient presentation of lipid antigens that traffic through the early endocytic and recycling pathways.

Amongst the CD1 family the CD1a, CD1b, and CD1c isoforms are included in the CD1 group 1 based on their sequence homology [Porcelli & Modlin 1999], are detected on activated human monocytes [Gregory et al. 2000; Sloma et al. 2004], and are highly expressed on monocyte-derived immature DC [Martin et al. 1986] and on dermal DC *in vivo*. During their maturation, DC undergo cellular changes which increase peptide loading and cell surface transport of major histocompatibility complex (MHC) class II-peptide complexes while reducing MHC class II endocytosis [Pierre et al. 1997; Turley et al. 2000]. An important difference between presentation of peptide and lipid antigens is that optimal peptide presentation to MHC class II-restricted T cells is dependent on this maturation, whereas both immature and mature DC optimally present lipid antigens to CD1-restricted T cells [Cao et al. 2002]. The T cell response to peptide antigens is initiated by recognition of MHC class II expressed at the surface of professional APC. This recognition is improved when MHC class II-peptides complexes are concentrated on the APC plasma membrane in small lipid-rich microdomains [Huby et al. 1999; Poloso & Roche 2004]. Although MHC class II organization in lipid rafts of various cell types is now well documented, this is not the case for CD1 molecules. Two studies have demonstrated the constitutive and restricted presence of the CD1d isoform in the lipid rafts of mouse cells [Lang et al. 2004; Park et al. 2005], whereas the localization of human CD1 molecules in specialized surface microdomains of professional APC has not been explored.

We have focused our attention on CD1a localization and organization in immature DC. This study [Sloma et al. 2008] investigates which proteins associate with CD1a, the possible intervention of tetraspanins, proteins acting as molecular adaptors, the CD1a distribution in specialized microdomains at the cell surface of immature DC, and the functional consequences of this organization on lipid antigen presentation. The results described in this study demonstrate that CD1a molecules are associated with invariant chain (Ii) in immature DC, that CD1a and Ii are recruited to cell surface lipid rafts, and that colocalization of CD1a and Ii is dependent on cholesterol-dependent lipid rafts. Moreover, we show that CD1a-induced expression in a HeLa-class II transactivator (CIITA) cell line led to an increased trafficking of Ii to the cell surface, and that silencing of Ii in immature DC induced significant CD1a accumulation at the cell surface. Finally, we show the role of immature DC membrane lipid rafts in T cell response to glycolipidic antigen presented by CD1a. The overall results of the present study suggest that Ii and lipid rafts have an impact on CD1a surface expression and CD1a-restricted T cell response.

The expression of CD1a, Ii, and human leukocyte antigen (HLA)-DR molecules on the plasma membrane of immature DC was analyzed on day 5 and 7 of their differentiation from monocytes. Analysis of cell surface and intracellular expression showed that CD1a was highly expressed at the cell surface and poorly detected in intracellular compartments of immature DC [Sloma et al. 2008]. There was a low level of Ii chain expression and it was predominantly intracellular. The HLA-DR molecules were clearly detected both at the cell surface and in intracellular compartments. To find the partner molecules associated to CD1a at the surface of immature DC, we performed immunoprecipitation experiments.

13.1. CD1a associates with Ii, HLA-DR and CD9

With CHAPS lysates, three major proteins around 33–35 and 24 kDa were coprecipitated specifically with CD1a. With regard to previous studies we speculated that the p33–35 proteins detected in the CD1a precipitates may be the 33 and 35 kDa isoforms of the Ii and/or the 35 kDa HLA-DRα-chain. The CD1a precipitates were analyzed by Western blot with Ii- or HLA-DR-specific mAb and relatively high levels of Ii were detected in CD1a precipitates as compared with the relatively low levels of HLA-DR molecules (see Figure 13.1 (A), p.109). These results demonstrate the association of CD1a and Ii in immature DC. To confirm this association at the cell surface, we performed sequential immunoprecipitation experiments. Ii was immunoprecipitated from surface biotinylated immature DC and eluted material was reimmunoprecipitated with CD1a- or Ii-specific mAb. Ii isoform at ≈ 33 kDa and specific Ii dimers at ≈ 66-80 kDa [Neumann et al. 2001] were visualized in Ii precipitates. Reciprocally, surface CD1a was precipitated from Ii-immunoprecipitates (see Figure 13.1 (B), p.109), thus demonstrating that CD1a and Ii are indeed associated at the cell surface of immature DC. This observation does not preclude that a third molecular partner could be involved in this association. The tetraspanin CD9 has been shown to be mostly expressed at the plasma membrane of human immature DC [Unternaehrer et al. 2007], and it can also interact with MHC class II at the surface of human monocytes [Zilber et al. 2005]. We thus explored the possibility that the p24 protein detected in CD1a precipitates could be CD9. After solubilization in CHAPS, many surface-biotin-labeled proteins were coprecipitated specifically with CD9 including a 49 and 33 kDa protein [Engering & Pieters 2001]. Western blotting of CD1a precipitates with the CD9 mAb Syb.1, identified the p24 protein as CD9 (see Figure 13.1 (A), p.109). Identical experiments performed on immature DC lyzed in Brij-98 led to the same observations (data not shown). These results show that CD1a associates with the Ii, the tetraspanin CD9 and MHC class II molecules, and demonstrate the CD1a/Ii association at the cell surface in human immature DC.

13.2. Ii silencing increases CD1a at the cell surface

While in steady-state, the four molecules are uniformly distributed at the cell surface, cross-linking of CD1a led to its redistribution either in caps or in patches at the cell surface [Sloma et al. 2008]. Dynamic recruitment of CD1a at the surface of immature DC was demonstrated to induce a highly significant redistribution of Ii and CD9 together with CD1a and partial redistribution of HLA-DR.

After sorting from the Golgi apparatus, Ii is mainly directed to endo-/lysosomal compartments. Nevertheless, a pool of Ii can also escape from this trafficking and directly reach the cell surface without entering endosomal compartments. As we have observed a tight link between CD1a and Ii at the surface, we asked whether CD1a could participate in cell surface transport of Ii. We have established HeLa-CIITA cell lines stably expressing CD1a or CD1b fused to green fluorescent protein (GFP), and controlled that GFP does not modify the cell surface or the cellular trafficking of the proteins as compared with wild type (WT) ones, as previously demonstrated [Manolova et al. 2006]. We have analyzed the surface and total expression of Ii and HLA-DR on these CD1a$^+$ cells and compared the results with those observed on CD1b$^+$ cells and mock-treated HeLa-CIITA cells. A significant increase in the surface expression level of Ii was observed in CD1a$^+$ HeLa-CIITA cells, while the total (surface and intracellular) Ii expression was unchanged in comparison with the non-CD1a expressing HeLa-CIITA cell lines [Sloma et al. 2008]. This indicates that, on CD1a$^+$ HeLa-CIITA cells, the proportion of Ii molecules transported to the surface is increased. Concerning HLA-DR expression, these cells harbored a slight and equivalent increase in both surface and total expression. This suggests a global increase of HLA-DR expression in those cells without any change in surface versus intracellular HLA-DR distribution.

The expression of Ii and MHC class II molecules in CD1b-expressing HeLa-CIITA cells was equivalent to that observed on mock cells: a slight decrease of Ii was observed at the cell surface as well as in permeabilized cells, and HLA-DR was decreased at the cell surface. Thus, the increase in Ii surface transport was a specific effect of the CD1a expression in HeLa-CIITA cells.

To further investigate the functional relevance of CD1a/Ii association on immature DC, we performed ribonucleic acid (RNA) interference experiments to knockdown the Ii expression on immature DC. Immature DC obtained at day 5 of monocyte differentiation were transiently transfected with small interfering RNA (siRNA) specific for Ii and analyzed by cytofluorimetry.

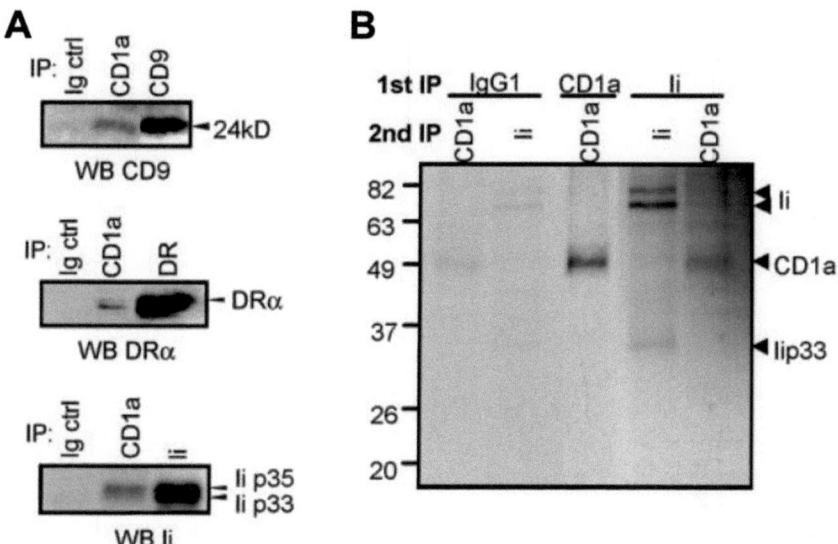

Figure 13.1
CD1a associates with CD9, HLA-DR and Ii. (A) Unlabeled immature DC were lyzed in CHAPS and CD1a precipitates were analyzed by Western blot with anti-CD9 (Syb.1), anti-HLA-DRα (DA6.147), or anti-Ii (LN2) mAb. (B) Ii molecules were immunoprecipitated from biotin-labeled immature DC. Material eluted from Ii precipitates was then subjected to a second immunoprecipitation with CD1a- and Ii-specific mAb. CD1a and isotype control mAb were used in first round of immunoprecipitation as controls: no material was detected in IgG1 negative control precipitates and CD1a was clearly detected in CD1a positive control precipitates. β_2 microglobulin (β_2m) had run off the bottom of the gel to concentrate attention on proteins with a molecular weight between 20 and 80 kDa. The data shown are representative of four independent experiments.

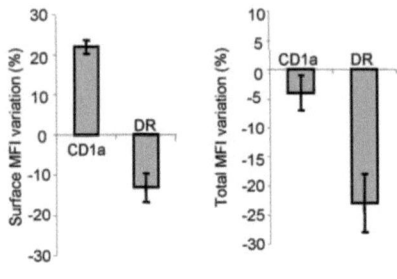

Figure 13.2
Ii interference induces CD1a accumulation at the cell surface of immature DC. Cell surface and total expression of CD1a (mAb HI149) and HLA-DR (mAb D1.12) on siRNAIi-treated immature DC expressed as percentage mean fluorescence intensity variation with respect to siRNA control-treated immature DC. Around 40 cells were examined in each preparation and the images shown are representative cells observed in the same preparation. The data shown are representative of four independent experiments.

Expression of CD83 was not detected on immature DC transfected with Ii- or control-siRNA indicating that transfection did not induce DC maturation [Sloma et al. 2008]. Under these conditions, the total expression of Ii was reduced by 95% (data not shown). As expected, the surface and total HLA-DR expression was decreased (see Figure 13.2, p.110). Silencing of Ii resulted in an increase of surface CD1a expression by ≈ 20% (see Figure 13.2, p.110). Therefore, these results indicate that in the absence of Ii, CD1a molecules accumulate at the cell surface and suggest that the Ii participates in CD1a endocytosis.

13.3. CD1a partition to detergent-resistant membrane microdomains is necessary for efficient exogenous antigen presentation

Clustering of proteins at the cell surface involves different actors and among them the tetraspanins have been implicated due to their capacity to link numerous proteins and raft microdomains. For this reason, we next asked whether there was a particular localization of these molecules in specialized membrane microdomains [Simons & Ikonen 1997; Xavier & Seed 1999]. Protein examination and quantification in Triton X-100 lysates of immature DC after fractionation by sucrose-gradient density ultracentrifugation revealed that a significant proportion (12.8%) of CD1 molecules is found in lipid rafts. To determine the localization of CD1, HLA-DR, Ii, and CD9 molecules in cell surface lipid rafts, co-patching experiments at the surface of living cells were performed using cholera toxin β, which binds to GM1, a constituent of cell surface lipid rafts. Z-stack analysis on single cells confirmed total and partial colocalization (data not shown) of Ii and CD1a with cholera toxin β, respectively. In agreement with previous studies [Poloso et al. 2006; Zilber et al. 2005], only a minor fraction of CD9 molecules colocalized with GM1. We then examined whether the dynamic recruitment of Ii and CD9 with CD1a at the surface of immature DC is dependent on lipid raft integrity. Cells were treated with methyl-β-cyclodextrin (MβCD) to deplete cholesterol and disrupt lipid rafts, and colocalization experiments were performed as described above. Induced CD1a relocalization after MβCD treatment no longer led to the enrichment of Ii into CD1a patches. The few areas of HLA-DR/CD1a colocalization observed before treatment with MβCD were almost completely lost by cholesterol depletion. In contrast, the CD1a/CD9 colocalization was still observed after raft disruption. These results indicate that the recruitment of Ii with CD1a is highly dependent on lipid raft integrity. To further investigate the CD1a/Ii association in immature DC surface lipid rafts, CD1a was pulled-down from sucrose gradient fractions of surface biotinylated immature DC and revealed surface protein

Figure 13.3

CD1a partitions to detergent-resistant membrane microdomains, which are necessary for efficient CD1a-restricted exogenous antigen presentation. CD14$^+$ monocytes from human donors were differentiated into immature monocyte-derived DC and cholesterol depleted by treatment with methyl-β-cyclodextrin. DC were then fixed and incubated with indicated concentrations of sonicated antigen and CD8-2 T cells, which recognize dideoxymycobactin presented by CD1a. After 21 h, the release of IFN-γ was measured by ELISA. Error bars represent SD of triplicate measurements, and the results are representative of three independent experiments.

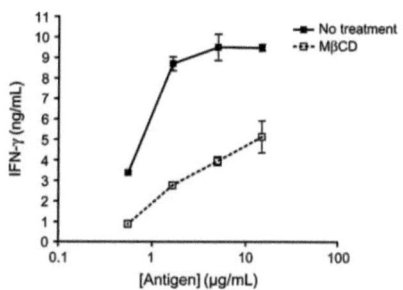

with streptavidin Western blotting. We observed that CD1a was equally distributed throughout the gradient, including the lipid raft enriched fractions, confirming that CD1a is present in cell surface lipid rafts of immature DC in steady state conditions. Moreover, a p33 protein corresponding to the molecular weight (MW) of Ii p33, was exclusively detected in CD1a precipitates from raft fractions (fraction 3–4) [Sloma et al. 2008]. Altogether these results demonstrate that the CD1a/Ii association is constitutive in immature DC and can take place during the dynamic recruitment of CD1a in lipid rafts.

To confirm that detergent-resistant membrane microdomains are important for CD1a-dependent antigen presentation, we tested the efficiency of presentation of a CD1a-dependent antigen in cholesterol-depleted immature monocyte-derived DC. We treated DC with MβCD, fixed the cells and incubated with dideoxymycobactin and CD8-2 T cells [Moody et al. 2004]. We found a striking decrease in the efficiency of presentation of this antigen, which in these experimental conditions has to be surface loaded (see Figure 13.3, p.111). The same was seen when MβCD-treated fixed immature DC were loaded with sulfatide and used to stimulate the CD1a-restricted sulfatide-specific human T cell clone K34B9.1 (data not shown). Non-fixed cells, in which MβCD is washed away after treatment, do not present any defect in dideoxymycobactin presentation (data not shown). Thus circumstances allowing for raft reconstitution restore the T cell response and this indicates that the cells are neither dead nor functionally impaired after cholesterol depletion. These results [Barral et al. 2008; Sloma et al. 2008] suggest that plasma membrane detergent-resistant membrane microdomains are important for CD1a-restricted antigen presentation in DC.

13.4. CD1a chimeras – tail to traffic twists

Albeit CD1 molecules are structurally similar to MHC class I, they functionally resemble MHC class II in regard to the ability of both molecules to survey endocytic compartments and, specifically, late endosomes (LE)/lysosomes for antigen acquisition.

The cytoplasmic tail of CD1b and CD1d [Barral & Brenner 2007] proteins has been shown to be critical for CD1 intracellular localization and its antigen-presenting function. The cytoplasmic tails of human CD1b, CD1c and CD1d, and also murine CD1d, all possess a tyrosine-based motif of the YXXϕ type in which Y is a tyrosine, X is any amino acid and ϕ is a bulky hydrophobic residue. In contrast, CD1a does not contain any recognizable sorting motifs (see Table 13.1,

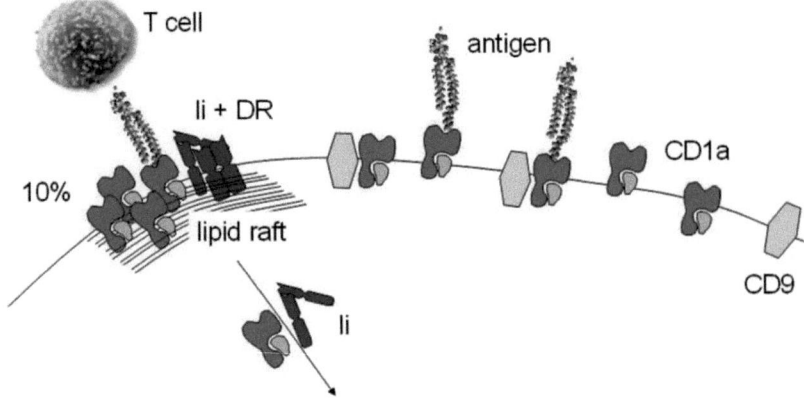

Figure 13.4
CD1a cartoon I - CD1a raft association, partners and dependency. Schematic view of a cell representing our findings that CD1a partitions (10-15%) into lipid rafts and associates with Ii, HLA-DR (DR), and CD9. CD1a/Ii interaction is dependent on lipid raft integrity whereas CD1a/CD9 interaction does not depend on lipid raft integrity. Ii is involved in CD1a internalization and recycling. Efficient activation of CD1a-restricted T cells is dependent on detergent-resistant membrane microdomains.

p.113). The deletion of the cytoplasmic tails of CD1b and murine CD1d reduces or abolishes the internalization and trafficking of the respective proteins to late endocytic compartments [Briken et al. 2002; Chiu et al. 2002; Jackman et al. 1998; Lawton et al. 2005; Sugita et al. 1996]. Moreover, tail-deleted (TD) CD1b and CD1d mutants display an impaired antigen-presenting capacity [Chiu et al. 1999; Chiu et al. 2002; Jackman et al. 1998]. In addition, some CD1 proteins possess putative dileucine-based sorting motifs [Moody 2006], although it has not been shown if these play any role in trafficking or function.

Table 13.1   CD1a chimeras – tail to traffic twists

molecule	tail [i]	trafficking
human CD1a	.RKRCFC	surface and sorting endosomes
human CD1a TD [ii]	.RKR	?
human CD1aab	.aTM [iii] & R(K)RRSYQNIP	?
human CD1abb	.bTM & RRRSYQNIP	?
human CD1b	.RRRSYQNIP	surface, LE [iv], lysosomes
human CD1c	.KKHCSYQDIL	surface and EE [v] (LE)
human CD1d	.KRQTSYQGVL	surface, LE, lysosomes
mouse CD1d	.RRRSAYQDIR	surface, LE, lysosomes

[i] Tyrosine-based motifs are in bold and red
[ii] Tail-deleted
[iii] Transmembrane
[iv] Late endosomes
[v] Early endosomes

Tyrosine- and dileucine-based motifs are known to bind adaptor protein (AP) complexes, such as AP-2, which is involved in endocytosis [Bonifacino & Traub 2003]. In fact, human CD1b and CD1c and murine CD1d cytoplasmic tails have been shown to bind AP-2 or one of its subunits [Briken et al. 2002; Lawton et al. 2005]. This last finding implies that CD1b, CD1c and CD1d are internalized in an AP-2-dependent manner, a pathway that is also mediated by clathrin, which coats the endocytic vesicles, and dynamin, which is thought to be involved in membrane fission. Indeed, CD1b internalization has been shown to be dynamin-dependent [Briken et al. 2002].

Despite lacking any known sorting motifs in its cytoplasmic tail, CD1a is internalized from the plasma membrane into endosomal compartments and recycles back to the plasma membrane through the endocytic recycling compartment [Salamero et al. 2001; Sugita et al. 1999]. It is not clear, however, if the CD1a cytoplasmic tail determines the intracellular trafficking pathways followed by this molecule. We investigated the internalization and recycling pathways followed by CD1a and the role of the CD1a cytoplasmic tail. Our results suggest that CD1a and MHC class I follow similar trafficking pathways within the endocytic system.

13.5. Tail-deletion does not change the subcellular distribution of CD1a

In a search of a possible role for the CD1a cytoplasmic tail in internalization and trafficking of the molecule, we analyzed the presence of posttranslational modifications that can occur in the cytoplasmic tail. Palmitoylation consists of the covalent attachment of a fatty acid (which can be palmitate or another fatty acid, hence the more appropriate designation S-acylation) to a cysteine. Contrary to myristoylation, another type of posttranslational lipid modification, there is no consensus sequence for S-acylation and it does not always occur at the N-terminus. Interestingly, CD1a possesses two cysteines in its cytoplasmic tail (see Table 13.1, p.113), which is striking considering that it only has six residues in total. We therefore investigated if CD1a could be S-acylated by incubating HeLa:CD1a stable transfectants with radiolabeled palmitic acid, followed by lysis and immunoprecipitation of CD1a [Barral et al. 2008]. As a control, we were able to detect labeled transferrin receptor (TfR), which is a known S-acylated protein. Strikingly, we also detected labeled CD1a, indicating that, indeed, CD1a is S-palmitoylated (data not shown). We predicted that, if CD1a is S-acylated in its cytoplasmic tail, a truncation of the tail would abolish the lipid modification. Therefore, we generated HeLa cells that stably express a CD1a tail-truncated mutant, with no cytoplasmic cysteines (HeLa:CD1a TD; (see Table 13.1, p.113)). Notably, we failed to detect labeled CD1a in HeLa:CD1a TD cells, indicating that CD1a is S-acylated on one or both cysteines of the cytoplasmic tail.

Many palmitoylated proteins, such as kinases and some Gα proteins [Resh 1999], are known to localize to detergent-resistant membrane microdomains or lipid rafts, which are domains enriched in cholesterol and sphingolipids [Magee & Parmryd 2003]. Also, glycosylphosphatidylinositol-anchored proteins, which follow a nonclathrin, dynamin-independent pathway of internalization [Sabharanjak et al. 2002], partition into detergent-resistant membrane microdomains establishing a dynamin-independent internalization pathway in these microdomains. As we had found that MHC class II Ii can associate with a fraction of CD1a molecules in immature DC and that this association is dependent on detergent-resistant membrane microdomains integrity where CD1a is partially located, we investigated if CD1a partitions into detergent-resistant membrane microdomains also in HeLa cells, which do not express Ii. We extracted the detergent-resistant membrane microdomains by lyzing HeLa:CD1a stable transfectants in cold Triton-X-100 and fractionating the lysate using a sucrose step gradient. Under these conditions, the detergent-resistant membrane microdomains float and can be isolated from other membranes. CD1a partitions together with a detergent-resistant membrane microdomains marker, caveolin, concentrating specifically in fraction number 10 (see Figure 13.5 (A,B), p.115). Nevertheless, most of the CD1a protein is detected in fractions that do not float (fractions numbers 1–4), although a significant amount of caveolin also partitions to these fractions. As a control, TfR, which is a non-detergent-resistant membrane microdomains protein, is virtually undetectable in detergent-resistant membrane microdomains-containing fractions [Barral et al. 2008]. These results suggest that, indeed, a fraction of the CD1a present in the cell is associated with detergent-resistant membrane microdomains, independently of Ii.

Because CD1a was shown to partition to detergent-resistant membrane microdomains and is S-acylated, we investigated if the CD1a TD mutant, which is not S-acylated, also partitions into detergent-resistant membrane microdomains. Interestingly, we were able to detect this mutant in detergent-resistant membrane microdomains fractions, suggesting that the S-acylation does not determine the partition of this protein into detergent-resistant membrane microdomains (see Figure 13.5 (C), p.115).

Figure 13.5

Tail-deletion does not change the subcellular distribution of CD1a. HeLa:CD1a WT (A) or HeLa:CD1a TD (C) stable transfectant cells were surface biotinylated, lyzed and fractionated in a sucrose step gradient. Fractions were collected from the bottom and CD1a immunoprecipitated (A and C). Samples were resolved on 12.5% SDS–PAGE and immunoblotted with HRP-conjugated streptavidin (A and C) or with rabbit polyclonal anti-caveolin (B), followed by HRP-conjugated anti-rabbit antibodies. In (A), associated β_2m is also shown. Results are representative of at least two independent experiments.

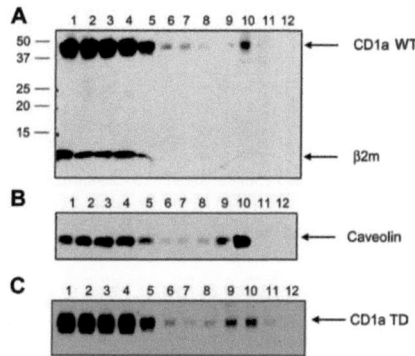

13.6. CD1a TD shows normal surface expression, internalization and recycling

S-acylation of the cytoplasmic tail of transmembrane proteins is known to influence their trafficking. For instance, S-acylation of one cysteine in the cytoplasmic tail of the cation-dependent mannose 6-phosphate receptor is critical for the correct trafficking and function of this receptor [Schweizer et al. 1996].

We therefore investigated whether the trafficking of CD1a TD is altered compared with CD1a WT. We first studied the expression of CD1a TD on the cell surface. We analyzed by flow cytometry the levels of CD1a on the surface of transiently transfected HeLa cells and saw no apparent difference in the surface expression of CD1a TD mutant when compared with CD1a WT (see Figure 13.6 (A), p.116). We also analyzed the internalization and recycling rates of CD1a TD. Surprisingly, the CD1a internalization rate was relatively fast and reached equilibrium after 20 min (see Figure 13.6 (B), p.116). This result was unexpected because CD1a does not have any sorting motif in its cytoplasmic tail, like other CD1 isoforms that possess tyrosine-based motifs (see Table 13.1, p.113), and it is not internalized in a clathrin-dependent manner (see Figure 13.8, p.119). Notably, when we compared the internalization rates of CD1a WT and CD1a TD, we saw no difference (see Figure 13.6 (B), p.116).

To study the recycling rates, we analyzed by flow cytometry the reappearance of CD1a on the cell surface after internalizing the protein bound to an antibody and stripping the antibody bound to non-internalized proteins. Similar to the internalization and surface expression studies, no differences between the recycling of CD1a WT and CD1a TD were detected (see Figure 13.6 (C), p.116). Therefore, these results identified no differences in trafficking between CD1a WT and CD1a TD. To confirm this, we compared the intracellular localization of CD1a WT and CD1a TD by transfecting these constructs into HeLa cells and performing confocal microscopy analysis. CD1a is known to recycle through the endocytic recycling compartment without trafficking to late endocytic compartments [Sugita et al. 1999]. Therefore, we used transferrin (Tf) ligand, which also recycles through the endocytic recycling compartment [Killisch et al. 1992], as a marker for this compartment. CD1a TD, similarly to CD1a WT, displays punctate staining after internalization and accumulation in the perinuclear region where the endocytic recycling compartment is localized. As expected, there is colocalization with Tf, suggesting that there are no differences in trafficking after internalization between the two molecules [Barral et al. 2008] at this level of resolution.

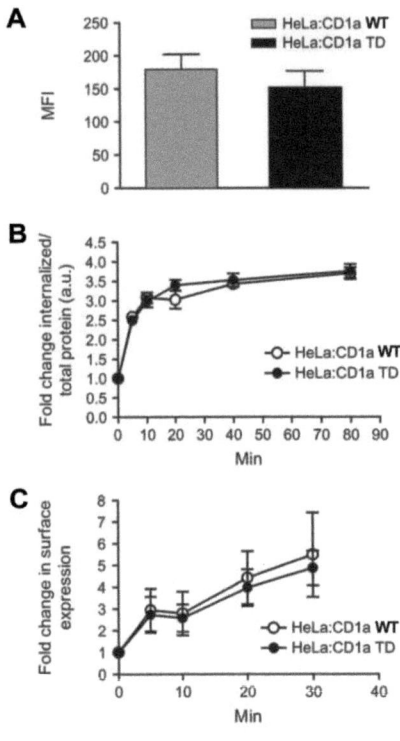

Figure 13.6

CD1a TD shows normal surface expression, internalization and recycling. (A) HeLa cells were transiently cotransfected with CD1a WT and enhanced GFP (EGFP) or CD1a TD and EGFP, stained for CD1a and analyzed by flow cytometry, gating on EGFP$^+$ cells. (B) HeLa:CD1a WT and HeLa:CD1a TD stable transfectants were surface biotinylated with a cleavable biotin at 4 °C, shifted to 37 °C for the indicated periods of time (to allow for internalization), after which cells were placed on ice to stop internalization. Surface biotin was cleaved with a reducing agent (on ice), and cells were then lyzed and lysates probed for different proteins by ELISA with each condition performed in duplicate. The results (in arbitrary units) are normalized for the total CD1a protein present in each lysate and the internalization at time zero. (C) HeLa cells were transiently cotransfected with CD1a WT and EGFP or CD1a TD and EGFP, stained for CD1a and incubated at 37 °C to allow for internalization. After stripping the surface-bound antibody with a brief acidic wash, cells were incubated at 37 °C for different periods of time to allow for recycling. Cells were then stained with phycoerythrin-labeled antimouse secondary antibody and analyzed by flow cytometry, gating on EGFP$^+$ cells. The results (in arbitrary units) are normalized for the recycling at time zero. Error bars in (A–C) represent the SD of three independent experiments.

Together, these results indicate that the S-acylation of the CD1a cytoplasmic tail does not play a significant role in the internalization or recycling of the protein or its localization to detergent-resistant membrane microdomains.

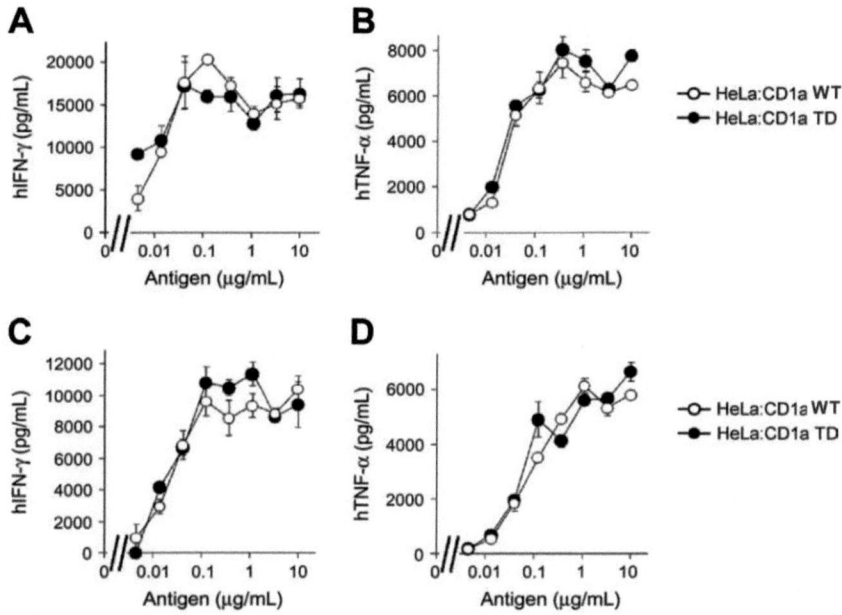

Figure 13.7
CD1a TD efficiently presents sulfatide to T cells. Nervonoyl sulfatide, at the indicated concentrations, was incubated with HeLa:CD1a WT or HeLa:CD1a TD and the CD1a-restricted sulfatide-specific human T cell clone K34B9.1 during the whole assay (A and B) or pulsed for 1 h at 37 °C (C and D). Supernatants were collected after 48 h, and release of IFN-γ (A and C) and TNF-α (B and D) was measured by ELISA. Error bars represent the SD of triplicate measurements.

13.7. Tail-deletion does not affect presentation of sulfatide to T cells

All the assays described thus far failed to show any differences in localization and trafficking between CD1a WT and CD1a TD. However, differences below the detection level of our assay systems could still lead to functional defects. We therefore compared the ability of the CD1a TD to present the antigen sulfatide with specific CD1a-restricted T cells [Shamshiev et al. 2002] and compared this with CD1a WT. When transfected HeLa cells were pulsed with sulfatide antigen, recognition by CD1a-restricted, sulfatide-specific T cells revealed no significant difference between CD1a WT and CD1a TD (see Figure 13.7, p.117). This suggests that the CD1a cytoplasmic tail does not appear to influence the antigen-presenting function of CD1a.

13.8. CD1a surface expression is not changed by inhibition of the clathrin pathway

CD1a was previously found in clathrin-coated pits and clathrin-coated vesicles [Salamero et al. 2001; Sugita et al. 1999]. However, unlike other CD1 isoforms, CD1a does not possess a tyrosine-based motif or other known sorting motifs that can bind AP-2 and mediate clathrin-dependent internalization (see Table 13.1, p.113). Therefore, we investigated the pathway of internalization followed by CD1a. We first utilized a knockdown approach directed at the clathrin heavy chain and the $\mu 2$ subunit of AP-2 using siRNA. For this, HeLa cells stably expressing CD1a or CD1b were transfected with siRNA oligos [Barral et al. 2008] and analyzed for the presence of CD1 molecules on the surface by flow cytometry (see Figure 13.8, p.119). As a control, TfR, a molecule known to follow a clathrin-mediated pathway of internalization, was used. The TfR showed a marked increase in surface expression when either the clathrin heavy chain or the $\mu 2$ subunit of AP-2 were knocked down. Strikingly, CD1a surface expression did not show any changes (see Figure 13.8 (A), p.119), contrary to CD1b [Barral et al. 2008] that is known to bind AP-2 and follow a clathrin-mediated internalization pathway [Briken et al. 2002]. We also disrupted the clathrin-mediated pathway by using a dominant-negative mutant of dynamin 2. HeLa cells stably transfected with CD1a or CD1b and transiently transfected with dynamin 2 K44A dominant-negative mutant (defective in GTP binding) expectedly showed an increase in TfR surface expression [Barral et al. 2008]. In accordance with the results obtained with siRNA, the surface expression of CD1a did not show significant changes, while CD1b surface levels were upregulated. These results suggest that CD1a follows a clathrin-, AP-2- and dynamin-independent internalization pathway. The experiments directly measuring CD1a internalization in dynamin 2 K44A transfectants were not fully conclusive as they showed a small but nonsignificant reduction (data not shown). Importantly, we also analyzed human immature monocyte-derived DC, which normally express all CD1 isoforms. We detected increased surface levels of TfR when AP-2 $\mu 2$ was knocked down by short hairpin RNA (shRNA) transduction, whereas no increase in CD1a surface levels could be detected (see Figure 13.8 (A), p.119). Moreover, we detected a decrease in CD1a surface expression, further suggesting that CD1a and TfR follow different pathways of internalization.

13.9. CD1a follows a Rab22a-dependent recycling pathway

As shown in (see Figure 13.8, p.119), the CD1a internalization pathway differs from the one followed by cargo internalized in a clathrin-dependent manner, such as TfR. Recently, the small guanosine triphosphatase Rab22a was shown to be involved in the recycling of proteins that are not dependent on clathrin for internalization [Weigert et al. 2004]. This recycling pathway, which can be distinguished from the classical slow recycling pathway followed by TfR, is inhibited by the expression of Rab22a constitutively active or dominant-negative mutants [Weigert et al. 2004]. However, a different study found significant inhibition in the recycling of TfR by a Rab22a constitutively active mutant [Magadan et al. 2006]. This disparity could be because of the difference in the species origin (human versus canine) of Rab22a used [Magadan et al. 2006]. We used the same human constructs, as used in the study mentioned first [Weigert et al. 2004], in HeLa cells and saw striking colocalization between CD1a and Rab22a (see Figure 13.9 (D), p.120) and CD1a and Rab22a-Q64L (defective in GTP hydrolysis) constitutively active mutant (see Figure 13.9 (H), p.120) in the tubular structures that represent tubular recycling endosomes [Weigert et al. 2004], suggesting that CD1a follows a Rab22a-dependent recycling pathway. Importantly, TfR could not be detected in these tubular recycling endosomes (see Figure 13.9 (C,F), p.120), consistent with the results obtained in the other study using the same

Figure 13.8

CD1a surface expression is not affected by inhibition of the clathrin-mediated pathway. (A) HeLa:CD1a stable transfectants were transiently transfected with clathrin heavy chain or μ2 subunit of AP-2 (μ2) siRNA oligos, twice at intervals of 72 h, then stained with anti-CD1a or anti-TfR antibodies and analyzed by flow cytometry. Error bars represent the SD of three independent experiments. For results with HeLa:CD1b see [Barral et al. 2008]. (B) CD14$^+$ monocytes were differentiated into immature monocyte-derived DC and transduced (or not) with GFP or μ2 subunit of AP-2 (μ2) shRNA encoding lentivirus for 6 days and analyzed in the same way. The MFI is represented in all cases, and the error bars represent the SD of four different shRNA for μ2.

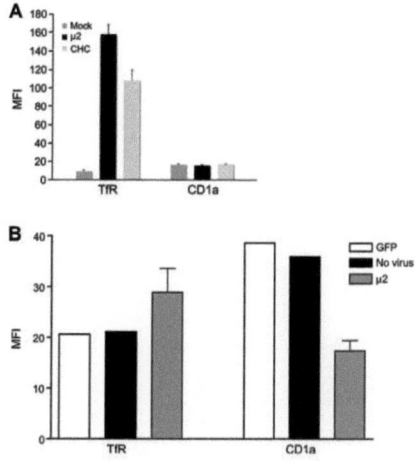

constructs [Weigert et al. 2004]. When we used a Rab22a dominant-negative mutant (Rab22a S22N, defective in GTP binding as is S19N), we detected a marked reduction in the CD1a recycling rate, strongly suggesting that CD1a follows a Rab22a-dependent recycling pathway (5.07-fold change in CD1a surface expression after 30 min in the control versus 2.43-fold change in cells transfected with the dominant-negative mutant Rab22a S22N). The residual recycling observed in the dominant-negative mutant could be because of an incomplete block of Rab22a-dependent recycling or the existence of an alternative recycling pathway.

Proteins of the CD1 family show similarity to MHC class I at the sequence and structural level. Interestingly, MHC class I also follows a clathrin- and dynamin-independent internalization pathway and a Rab22a-dependent recycling pathway [Naslavsky et al. 2003; Weigert et al. 2004], similar to what we observed with CD1a. Therefore, we determined if MHC class I and CD1a colocalize under steady-state conditions. We observed striking colocalization [Barral et al. 2008] between these two proteins, including in the perinuclear area. These data further suggest that CD1a and MHC class I follow similar intracellular trafficking pathways.

13.10. CD1a is internalized and recycled by an adenosine diphosphate ribosylation factor 6-dependent pathway

CD1a endosomal recycling has been shown to be adenosine diphosphate ribosylation factor (ARF) 6-dependent [Sugita et al. 1999]. Interestingly, MHC class I is also a marker for this pathway [Radhakrishna & Donaldson 1997]. We therefore made use of an ARF6 constitutively active mutant (ARF6-Q67L, defective in GTP hydrolysis), which blocks membrane trafficking shortly after internalization and sequesters cargo that normally traffics through the ARF6-dependent endocytic recycling pathway [Brown et al. 2001], such as MHC class I. HeLa cells transfected with the ARF6-Q67L mutant showed typical enlarged vacuolar structures where MHC class I accumulated (see Figure 13.10, p.121). Strikingly, CD1a, but not TfR (which is internalized in a clathrin-dependent manner), accumulated in the same structures.

119

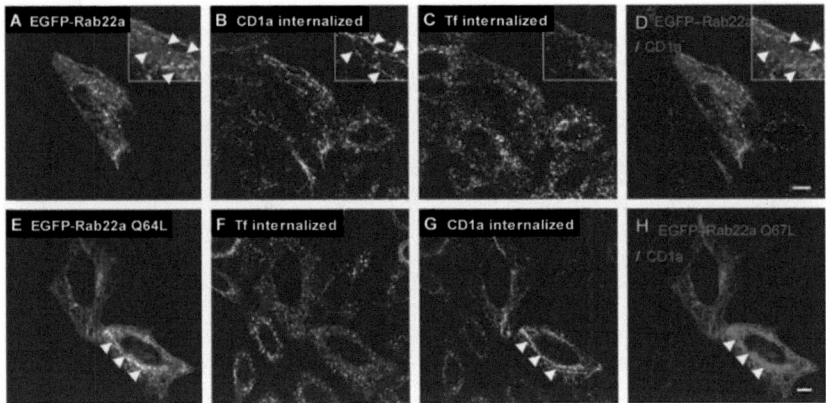

Figure 13.9

CD1a follows a Rab22a-dependent recycling pathway. HeLa:CD1a stable transfectants were transiently transfected with enhanced green fluorescent protein (EGFP)–Rab22a (WT or Q64L constitutively active mutant). After 24 h, cells were serum starved for 30 min and incubated with Alexa 647-conjugated Tf (C) or Alexa 546-conjugated Tf (F) and anti-CD1a mAb (B and G) for 30 min on ice. Cells were then washed and incubated with Alexa 647-conjugated Tf for 45 min at 37 °C (upper panels) or switched only to 37 °C for 30 min (lower panels). After stripping the surface-bound antibody with a brief acidic wash, cells were fixed, permeabilized and labeled with Alexa 546-conjugated anti-mouse (B) or Cy5-conjugated anti-mouse (G) antibodies. Arrowheads depict tubular structures that contain both EGFP–Rab22a and CD1a. In the merge panel (D), colocalization is indicated by the yellow color, while in panel (H), it is indicated by the color cyan. Scale bars, 10 μm.

13.11. CD1b does not accumulate in ARF6-Q67L-positive enlarged vesicles

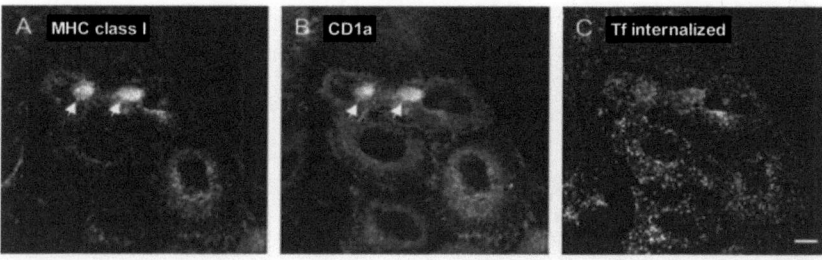

Figure 13.10
CD1a is internalized and recycled by an ARF6-dependent pathway. HeLa:CD1a stable transfectants were transiently transfected with ARF6-Q67L constitutively active mutant. After 24 h, cells were serum starved for 30 min and incubated with Alexa 647-conjugated Tf (C) for 30 min on ice. Cells were washed, incubated with Alexa 647-conjugated Tf for 30 min at 37 °C and then fixed, permeabilized and incubated with Alexa 488-conjugated anti-MHC class I (A) and Alexa 546-conjugated anti-CD1a (B) antibodies. Arrowheads depict typical enlarged vacuolar structures induced by ARF6-Q67L expression. Scale bar, 10 µm.

13.11. CD1b does not accumulate in ARF6-Q67L-positive enlarged vesicles

As expected, CD1b, which is internalized in an AP-2- and dynamin-dependent manner [Briken et al. 2002], could not be detected in these enlarged vacuolar structures (see Figure 13.11, p.122). These results are in agreement with the finding that CD1a follows an AP-2- and dynamin-independent internalization pathway (see Figure 13.8, p.119) and strongly suggest that CD1a follows a similar recycling pathway to MHC class I. Furthermore, they indicate that these two proteins not only share structural similarities but also are similar in their intracellular trafficking.

13.12. Relocation of CD1a to lysosomes by providing residues of CD1b cytoplasmic tail

It has been previously shown that CD1a localizes to the endocytic recycling compartment, whereas, CD1b localizes to LE or lysosomes. To investigate the importance of tyrosine-based motifs in CD1 molecule trafficking, HeLa CD1a:CD1b tail chimeric constructs (see [Cernadas et al. 2009] and references therein) containing the tyrosine-based motif of CD1b were generated (see Table 13.1, p.113) and intracellular localization of CD1 molecules analyzed by confocal microscopy. The CD1aab tail chimera showed a different localization from CD1a WT and an identical staining pattern compared with WT CD1b (see Figure 13.13 (A), p.124). This was further demonstrated by colocalization of the CD1aab chimera to LE/lysosomes with lysosome-associated membrane glycoprotein 1 (LAMP-1) as it was also demonstrated for WT CD1b. This is in agreement with the known important role for the tyrosine-based sorting motif of CD1b in binding AP-3 and directing trafficking to the lysosomes. In contrast and as expected, there was no colocalization between WT CD1a and LAMP-1 (see Figure 13.13 (B), p.124). Both WT and chimeric proteins were well expressed on the cell surface as demonstrated by flow cytometry. The CD1aab chimera transfectants consistently showed lower surface mean fluorescence inten-

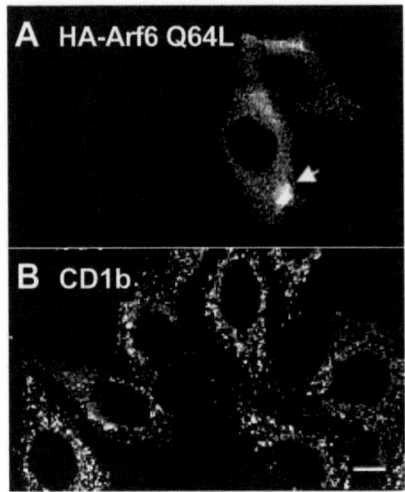

Figure 13.11

CD1b does not accumulate in ARF6-Q67L-positive enlarged vesicles. HeLa:CD1b stable transfectants were transiently transfected with ARF6-Q67L constitutively active mutant. After 24 h, cells were fixed, permeabilized and incubated with anti-hemagglutinin polyclonal antibody (A) and anti-CD1b mAb BCD1b3.2 (B), followed by Alexa 488-conjugated anti-rabbit and Alexa 546-conjugated anti-mouse secondary antibodies. Arrowhead depicts an enlarged vacuole induced by ARF6-Q67L expression. Scale bar, 10 µm.

sity compared to WT CD1a as determined by flow cytometry (data not shown). It should be noted that there was significant anti-CD1a mAb intracellular staining with anti-CD1a mAb of the CD1aab chimera indicating conformational stability of the CD1a extracellular domain in the late endocytic system (see Figure 13.13 (A), p.124).

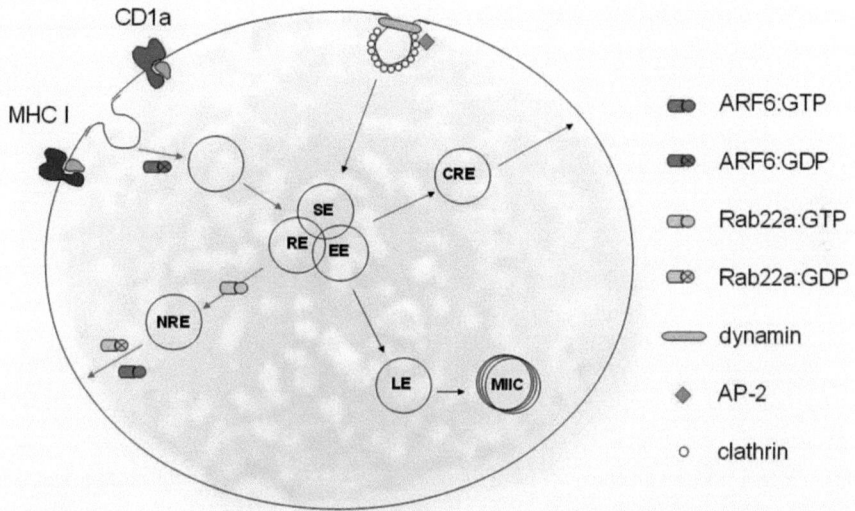

Figure 13.12
CD1a cartoon II - CD1a mimics MHC class I trafficking. Scheme of a cell depicting our results that the CD1a pathway does not require any motifs in the cytoplasmic tail but follows a similar intracellular trafficking route as MHC class I (MHC I). Recycling is dependent on ARF6 and Rab22a and internalization is neither dependent on clathrin, nor AP-2, nor dynamin. Nevertheless, CD1a shows fast internalization kinetics by reaching the steady state equilibrium after 20 min. Mutants studied were dynamin 2 K44A (dominant negative mutant defective in GTP binding), Rab22a S19/22N (dominant negative mutant defective in GTP binding), Rab22a Q64L (constitutively active mutant defective in GTP hydrolysis), ARF6 T27N (dominant negative mutant defective in GTP binding) and ARF6 Q67L (constitutively active mutant defective in GTP hydrolysis). Routes are abbreviated as follows: CRE = clathrin recycling endosomes, EE = early endosomes, LE = late endosomes, MIIC = MHC class II compartment, NRE = nonclathrin recycling endosomes, RE = recycling endosomes and SE = sorting endosomes. Red arrows indicate the routes taken by MHC class I and CD1a.

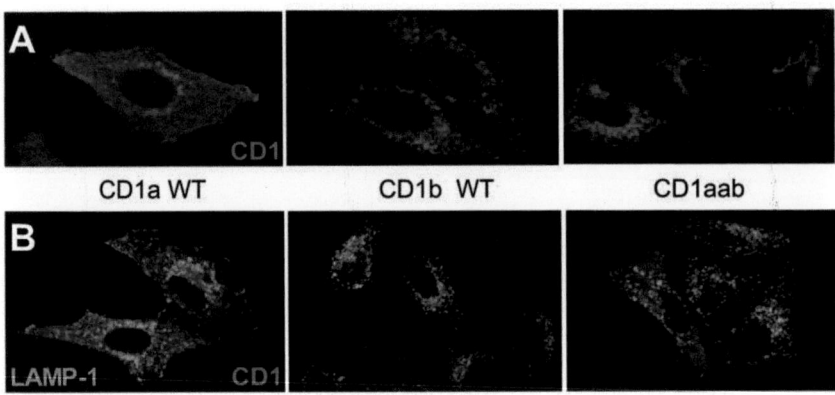

Figure 13.13
CD1a is redirected to lysosomes by providing residues of CD1b cytoplasmic tail. HeLa CD1a WT, CD1b WT and CD1aab tail chimera transfectants were analyzed by confocal microscopy. (A) The CD1 transfectants were stained with antibodies directed at their extracellular domains. CD1a WT transfectants show cell surface and perinuclear staining. The CD1b WT and CD1aab chimera have similar intracellular staining patterns. (B) Colocalization studies with anti-LAMP-1 mAb demonstrated the localization of the CD1aab chimera to LAMP-1^+ compartments similar to WT CD1b. There is no colocalization between CD1a and LAMP-1. The fluorescent label used for each protein is indicated by color.

13.13. Colocalization of CD1a and sulfatide

Several T cell antigens presented by CD1a have been described including dideoxymycobactin, a lipopeptidic antigen isolated from *Mycobacterium tuberculosis*, and sulfatide [Moody et al. 2004; Shamshiev et al. 2002]. Sulfatide (3'-sulfated β1-D-galactosylceramide) is an endogenous glycolipid highly expressed in neuronal cells, kidney and pancreas [Fredman et al. 2000; Sandhoff et al. 2002; Vos et al. 1994] and its synthesis is upregulated in DC upon bacterial infection [De Libero et al. 2005 a]. Sulfatide can be presented by all group 1 CD1 molecules including mouse and human CD1d [Shamshiev et al. 2002; Zajonc et al. 2005 a] (M. Cavallari and G. De Libero, unpublished results). It can be efficiently loaded on the cell surface of DC and can be presented without internalization or endosomal acidification [Shamshiev et al. 2002]. Differences were observed in the antigen presentation capabilities of sulfatide-pulsed DC for each of the CD1 isoforms over time. DC pulsed with sulfatide maintained the ability of DC to stimulate CD1a-restricted T cells over a three day period. In contrast, the ability of the DC to stimulate CD1b- and CD1c-restricted T cells was reduced by $\approx$ 75% over the same time period. These findings suggest an especially prolonged interaction between CD1a and sulfatide in myeloid DC. However, these studies cannot distinguish between higher affinity of CD1a binding for sulfatide versus more extensive colocalization of the antigen presenting molecule with the antigen.

Despite extensive prior analyses of CD1a and CD1b trafficking, only rarely has lipid antigen trafficking been directly visualized, since tagging of lipid antigens often alters their trafficking and few mAb against lipids exist. No prior studies have examined the trafficking of CD1a-presented lipid antigens. We developed a protocol to determine the intracellular distribution of sulfatide using confocal microscopy and the anti-sulfatide specific mAb (O4) in sulfatide-pulsed HeLa CD1 transfectant cells. Strikingly, HeLa CD1a transfectants incubated with sulfatide demonstrated an almost exclusive localization of sulfatide to the early endocytic and recycling pathways [Cernadas et al. 2009], identified using anti-Rab5- and early endosome antigen 1-specific mAb and fluorescently-labeled Tf. There was marked colocalization between sulfatide and the early endocytic markers Rab5 and early endosome antigen 1. Sulfatide was also present in the endocytic recycling compartment as demonstrated by its partial colocalization with Tf. CD1a and Tf also demonstrated partial overlap consistent with the previously published [Barral et al. 2008; Sugita et al. 1999] localization of CD1a in the endocytic recycling compartment. However, there was no colocalization with LAMP-1, a marker of the late endocytic system.

Next, the colocalization pattern of CD1a and CD1b and the CD1a:CD1b chimera with sulfatide was examined. HeLa CD1a WT, CD1b WT and CD1a:CD1b chimera transfectants were pulsed with sulfatide and stained with anti-CD1a, anti-CD1b and anti-sulfatide mAb. Consistent with the above findings there was a marked colocalization of CD1a with sulfatide in the HeLa CD1a WT transfectant cells (see Figure 13.14 (A), p.126). On the other hand, little to no colocalization of CD1b with sulfatide was found in the HeLa CD1b WT transfectants (see Figure 13.14 (A), p.126). Compared to CD1a WT, there was a marked reduction in the colocalization of sulfatide with the CD1aab tail chimeric protein (see Figure 13.14 (A), p.126). This suggests that the intracellular localization provided by the CD1b tyrosine-based motif determines the degree of interaction of the CD1aab tail chimera with sulfatide. The distribution of sulfatide was very similar in cells expressing CD1a WT and CD1aab tail chimera as demonstrated by the almost complete colocalization of sulfatide and Rab5 and there was no redirection of sulfatide to the late endocytic system with the CD1aab tail chimera (see Figure 13.14 (B), p.126).

Time course studies of sulfatide distribution to investigate whether sulfatide traffics to LE at late time points showed that there is no detectable colocalization of LAMP-1 with sulfatide [Cernadas et al. 2009] up to 20 h after sulfatide pulse. This lack of colocalization was observed for both the CD1a WT and CD1aab tail chimera transfectants. Minor colocalization was observed for sulfatide and CD1b at late time points (data not shown). At all time points the vast majority

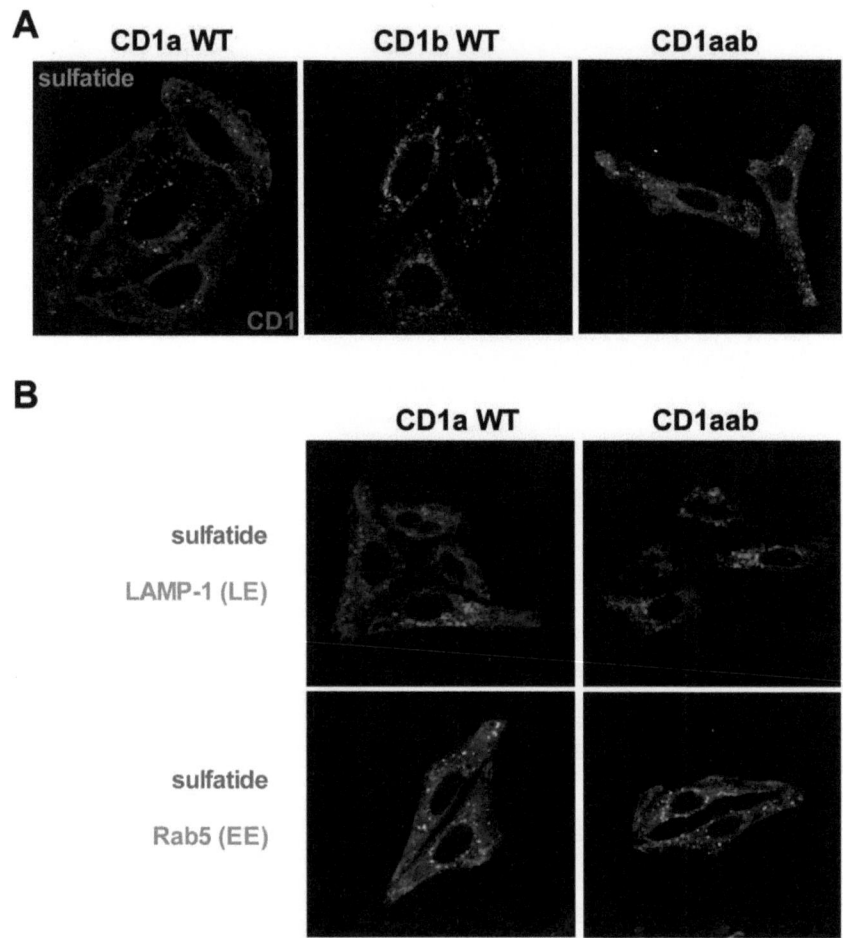

Figure 13.14
Marked intracellular colocalization between CD1a and sulfatide. HeLa CD1a WT, CD1b WT and CD1aab chimera transfectants were pulsed with sulfatide and analyzed by confocal microscopy. (A) Colocalization between each CD1 molecule and sulfatide was determined for each transfectant. The fluorescent labels used for each protein are indicated by color. (B) Colocalization between sulfatide and early (Rab5) and late endocytic (LAMP-1) markers was determined in sulfatide-pulsed CD1a WT and CD1aab tail chimera transfectants.

of sulfatide staining remained in the early endocytic and recycling compartments. The absence of sulfatide detection in LAMP-1$^+$ compartments even at later time points is likely due to its preferential sorting to the early endocytic and recycling compartments and to the degradation of sulfatide by endogenous lysosomal arylsulfatases that are well described to be present in this compartment [Diez-Roux & Ballabio 2005]. Lastly, to rule out the possible contribution of sulfatide to changes in CD1a expression, intracellular distribution or cell morphology, confocal microscopy was performed with increasing concentrations of sulfatide or ceramide (C24) as an additional control. No changes were observed by confocal microscopy in the distribution or localization of CD1a nor in the transfectant morphology [Cernadas et al. 2009] after culture with either lipid.

These findings [Cernadas et al. 2009] were confirmed in primary cells, namely human monocyte-derived DC that express all CD1 isoforms, and are in agreement with the known localization of CD1a and CD1b molecules in monocyte-derived DC to the endocytic recycling compartment or to late endosomal/lysosomal compartments, respectively [Cao et al. 2002; van der Wel et al. 2003]. Briefly, human monocyte-derived DC were pulsed with sulfatide as described above. Confocal microscopy of DC demonstrated areas of strong colocalization between sulfatide and CD1a molecules but not between sulfatide and CD1b.

In contrast to the specific colocalization of CD1a with sulfatide, lipoarabinomannan, an antigen presented by CD1b, strictly colocalized with CD1b but not with CD1a [Cernadas et al. 2009] in monocyte-derived DC. The localization of lipoarabinomannan with CD1b is consistent with the known interaction of lipoarabinomannan with mannose receptors that likely mediate the internalization and direct targeting of lipoarabinomannan and similar antigens to the late endocytic system [Schlesinger et al. 1996]. These findings suggest that CD1 molecules select the lipid antigen to be presented according to their capacity to bind and form stable complexes and also according to the endosomal localization of each lipid antigen and its intersection with the relevant CD1 molecule.

13.14. WT CD1a transfectants present sulfatide more efficiently than CD1aab chimera

In order to determine the functional relevance of the intracellular trafficking of CD1a in antigen presentation, we next examined the ability of CD1a WT and different CD1a:CD1b tail chimeras to activate CD1a-restricted sulfatide-specific T cells. First, to assess the appearance of sulfatide:CD1a stimulatory complexes we performed pulse experiments using HeLa CD1a WT and CD1aab chimera transfectants with matched cell surface expression of CD1a as determined by flow cytometry. The transfectants were pulsed with sulfatide for different time periods before addition of T cells. The ability of sulfatide-pulsed transfectants to stimulate the sulfatide-specific T cell clone K34B9.1 increased with increasing pulse period independently of the measured cytokine (IFN-γ, TNF-α or IL-4). CD1a WT transfectants outperformed the CD1aab chimera at pulse periods longer than 30 min (see Figure 13.15, p.128). CD1aab chimera transfectants even showed a slight decrease in stimulatory capacity at pulse periods longer than 30 min.

13.15. CD1a:CD1b chimeras have shorter-lived ability to stimulate T cells

Secondly, to assess the disappearance of sulfatide:CD1a stimulatory complexes we performed chase experiments using T2 CD1a WT and CD1a:CD1b chimeras instead of HeLa transfectants to avoid effects on stimulation due to trypsinization. The T2 transfectants were pulsed with

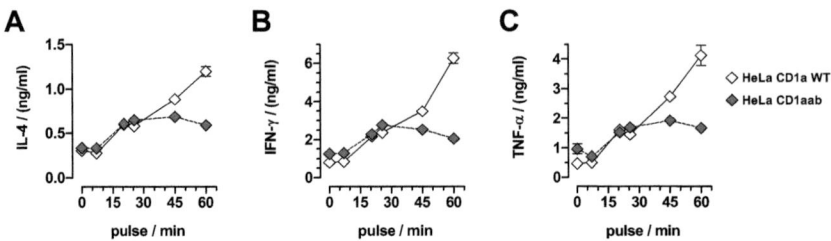

Figure 13.15
WT CD1a transfectants present sulfatide more efficiently than CD1aab chimeric transfectants.
WT CD1a (white diamonds) HeLa transfectants showed increased capacity to stimulate the CD1a-restricted sulfatide-specific T cell clone K34B9.1 as compared to CD1aab chimeric transfectants (gray diamonds) for pulse periods longer than 30 min. Both transfectants were pulsed with nervonoyl sulfatide for indicated times, then washed and co-incubated with T cells. Supernatants were harvested after 24 h and the release of IL-4 (A), IFN-γ (B) and TNF-α (C) was measured by ELISA. Error bars represent SD of triplicate measurements.

sulfatide and chased for indicated time periods before addition of T cells and cytokine release measurement. At each time point control groups were represented by transfectants incubated with sulfatide and T cells together to determine the maximal antigen presenting capacity. The ability of sulfatide-pulsed transfectants to stimulate the sulfatide-specific clone K34B9.1 decayed with increasing chase period independently of the measured cytokine. WT CD1a transfectants showed half maximal stimulation at 40 h of chase, whereas CD1abb transfectants showed half maximal stimulation at 20 h. CD1aab transfectants showed intermediate behavior, with half maximal stimulation at 30 h (see Figure 13.16, p.129). After 72 h CD1a WT transfectants were still capable of stimulating sulfatide-specific T cells, whereas both chimera transfectants were no longer stimulatory.

13.16. Conclusions and outlook

Our first study [Sloma et al. 2008] demonstrates that in human immature DC CD1a associates with the Ii, HLA-DR, and the tetraspanin CD9 and partitions into lipid rafts. The CD1a/Ii association takes place during recruitment of CD1a into surface lipid rafts and this type of membrane localization promotes efficient antigen presentation and T cell activation. We also provide evidence that on the one hand CD1a participates in controlling the pool of surface Ii and that on the other hand Ii participates in CD1a endocytosis. Our biochemical analysis revealed that several molecules, such as HLA-DR, Ii p35, Ii p33, and CD9 coprecipitate with CD1a. The amounts of Ii and CD9 proteins detected in CD1a precipitates are strikingly more abundant than those of coprecipitated HLA-DR. Cocapping experiments are in agreement with this finding, because confocal microscopy showed a faint presence of HLA-DR within CD1a clusters, and a prominent colocalization of CD1a with Ii and CD9. The scarce number of CD1a-HLA-DR clusters argues against their physiological relevance. The association of the CD9 tetraspanin with CD1a, which resembles the CD9/HLA-DR clusters present on human monocytes [Zilber et al. 2005], appears to be more important. Another CD1 isoform, CD1d, has been found associated with the CD82 tetraspanin and MHC class II molecules. CD1d may associate with MHC class II indepen-

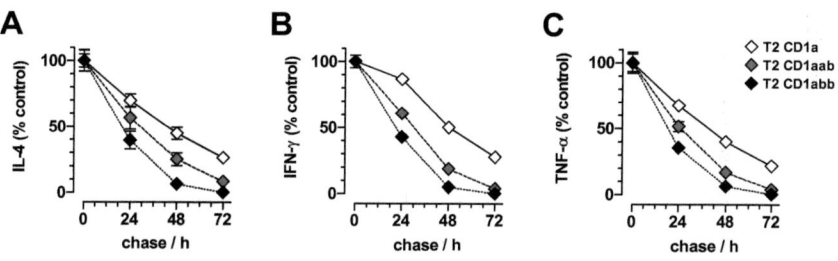

Figure 13.16
CD1a:CD1b chimeras have shorter-lived ability to stimulate sulfatide-specific T cells than WT CD1a molecules. WT CD1a (white diamonds) T2 transfectants showed prolonged capacity to stimulate the CD1a-restricted sulfatide-specific human T cell clone K34B9.1 as compared to CD1aab (gray diamonds) or CD1abb (black diamonds). All transfectants were pulsed with nervonoyl sulfatide and chased for indicated times before addition of T cells. In control wells fresh antigen was added to chased T2 cells together with T cells. Supernatants were harvested after 24 h and the release of IL-4 (A), IFN-γ (B) and TNF-α (C) was measured by ELISA and expressed as percent of control at time zero ± SD (of triplicates or quadruplicates) normalized at each time point for the maximal antigen presenting capacity of the APC calculated as $\frac{chase\ /\ chase\ with\ fresh\ antigen}{no\ chase\ /\ no\ chase\ fresh\ antigen} \times 100$.

dently of CD82 [Kang & Cresswell 2002 b]. The reason why CD1d does not coprecipitate with tetraspanins remains unclear. Tetraspanins associate with numerous surface molecules and form networks which organize the local distribution of interacting molecules. These networks are also formed in membrane domains outside lipid rafts, thus demonstrating the important general role of tetraspanins in membrane organization [Poloso et al. 2006; Tarrant et al. 2003]. The presence of large amounts of CD9 molecules that laterally associate with CD1a, may have the double effect of organizing CD1a in the membrane and enhancing CD1a-mediated antigen presentation, as has been reported for HLA-DR in DC [Jayawardena-Wolf et al. 2001; Unternaehrer et al. 2007]. A second mechanism used by proteins to concentrate at the cell surface is to enter lipid rafts microdomains [Xavier & Seed 1999].

Our results reveal that CD1a molecules are constitutively present in raft compartments of immature DC total lysates and in cell surface lipid rafts. A significant portion of Ii and MHC class II molecules that are associated with CD1a distribute in lipid rafts. Previous studies have described the association of MHC class II with lipid rafts in monocyte-derived human immature DC and have shown that only 10–20% of the total MHC class II pool localizes in rafts, the remaining MHC class II molecules being constitutively distributed in non-raft compartments of immature DC [Eren et al. 2006; Setterblad et al. 2003]. Similarly to MHC class II, we find that only 10–15% of CD1a molecules are detected in immature DC lipid rafts. To evaluate the impact of raft integrity on the CD1a proximity to Ii and CD9, we performed cholesterol depletion experiments. We found that CD1a/Ii colocalization is, to a great extent, dependent on the integrity of lipid rafts, whereas CD1a/CD9 colocalization is not. This observation can result from physical differences between lipid rafts and tetraspanin-enriched microdomains. Despite the potential of tetraspanins to associate with gangliosides and cholesterol, which typically are resident in lipid rafts, the tetraspanin-enriched microdomains are resistant to cholesterol depletion [Hemler 2003; Zilber et al. 2005]. The association of CD1a with Ii and the dynamic recruitment of these

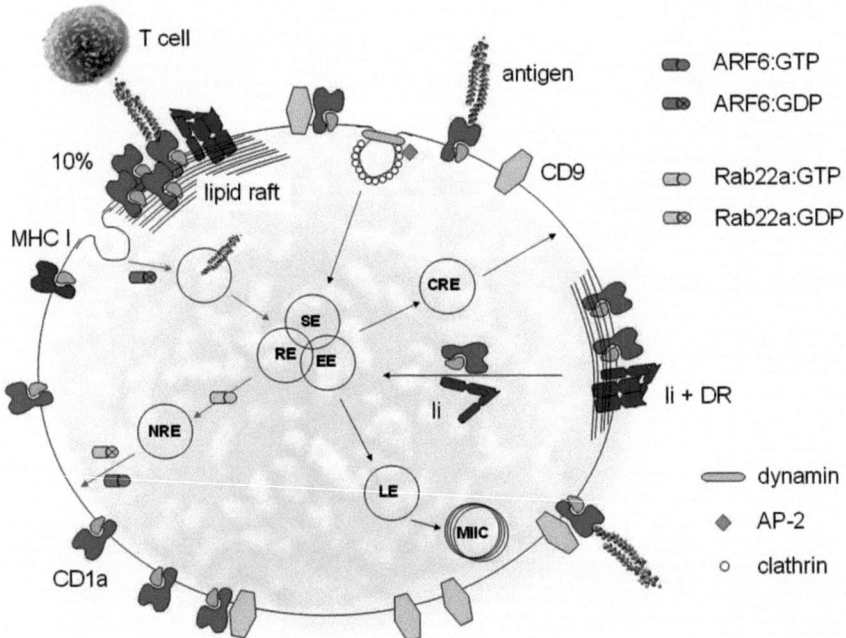

Figure 13.17
CD1a cartoon III - the view of CD1a trafficking and antigen-presentation to date. Presenting (see Figure 13.4, p.112) and cargo-associated molecules (see Figure 13.12, p.123) are shown as before. Routes are given as follows: CRE = clathrin recycling endosomes, EE = early endosomes, LE = late endosomes, MIIC = MHC class II compartment, NRE = nonclathrin recycling endosomes, RE = recycling endosomes and SE = sorting endosomes. The routes taken by MHC class I and CD1a are colored red.

complexes in lipid rafts has important functional consequences on CD1a antigen presentation. Indeed, CD1a association with lipid rafts facilitates activation of CD1a-restricted T cells, especially at low antigen doses. This result is in agreement with the finding that lipid raft recruitment of MHC class II molecules promotes the response of peptide-specific T cells [Anderson et al. 2000].

That study and our findings make the important point that the membrane localization of the presenting molecule is crucial independently of the type of presented antigen and responding T cell. It is tempting to speculate that clustering of antigen-loaded CD1a molecules has a major impact in the presence of low antigen amounts because it can facilitate TCR focusing and signaling cascade in T cells despite the low numbers of CD1a-antigen complexes. Another important question raised in this study is the role of Ii in CD1a trafficking to the plasma membrane and through the endocytic pathway. $\alpha\beta$Ii complexes may use two distinct trafficking pathways [Dugast et al. 2005]. The first pathway is dependent on AP-1, which promotes direct traffic to endosomes. The second, which is the major pathway, is dependent on AP-2, which allows rapid passage of these complexes to the cell surface. Therefore, Ii may impart different traffic behavior to associated proteins, according to the adaptor protein it is associated with. Ii has been implicated in direct membrane trafficking of MHC class II. For example, in CD1a-negative APC, such as B cells and monocytes [Delia et al. 1988; Gogolak et al. 2007] a minor fraction of newly synthesized MHC class II molecules can directly reach the cell surface as $\alpha\beta$/Ii-p33 complexes [Warmerdam et al. 1996]. Our coprecipitation data together with the increased Ii surface expression in the presence of CD1a suggest that a fraction of Ii molecules directly traffic to the plasma membrane when associated with CD1a. This mechanism could operate in immature DC, a cell population expressing large amounts of CD1a, in which the majority of newly formed $\alpha\beta$Ii complexes appear firstly on the cell surface before they start recycling [Saudrais et al. 1998]. However, the association with Ii is not an absolute requirement for surface appearance, because CD1a directly reaches the cell surface after egress from the endoplasmic reticulum (ER) and trans Golgi network in both Ii-positive or Ii-negative cells [Sugita et al. 1999]. Our experiments also show that extinction of Ii in primary human DC induces an increased surface expression of CD1a, which we interpret as a result of diminished internalization. Thus, Ii is not required for surface display of newly synthesized CD1a molecules, but might be involved in CD1a internalization and recycling. Endocytosis of other CD1 molecules is largely dependent on intracytoplasmic tail motifs allowing interaction with AP-2 [Briken et al. 2000]. The inability of the CD1a cytoplasmic tail to interact with AP-2 raises the question of the mechanism involved in CD1a internalization. We propose that CD1a molecules that associate with Ii benefit from the efficient internalization signal in the Ii cytoplasmic tail and then follow a fast internalization pathway. As a large pool of CD1a molecules binds exogenous ligands present in the extracellular environment on the cell surface [Manolova et al. 2006], it is possible that only the minor fraction of CD1a molecules associated with Ii recycle in clathrin-coated vesicles and early endosomes (EE) and sample intracellular antigens. Further investigations are needed to examine the physiological significance of Ii in CD1a endocytosis and presentation of intracellular antigens.

Our second study [Barral et al. 2008] focuses on trafficking of CD1a and its dependency on the cytoplasmic tail. Most cytoplasmic tails from different CD1 isoforms across species have well-characterized tyrosine-based sorting motifs or, alternatively, putative dileucine-based sorting motifs [Moody 2006]. Primate (including human) CD1a cytoplasmic tails are the shortest known with only three residues downstream of the three basic residues present in all CD1 proteins [Moody 2006] (C. C. Dascher, Mt. Sinai School of Medicine, New York, unpublished data). We investigated possible lipid modifications of the cysteines of the CD1a cytoplasmic tail and showed that the CD1a cytoplasmic tail is S-acylated on one or both cysteines. Interestingly, we were also able to detect S-acylation of CD1c, which has a cysteine next to the third residue of the cytoplasmic tail like CD1a, but not of CD1b, which does not have any cysteines in its cytoplasmic tail (D. C. Barral unpublished data) (see Table 13.1, p.113).

Protein S-acylation is known to be important for membrane targeting, localization to membrane microdomains, signaling and protein trafficking [Bijlmakers & Marsh 2003]. It is therefore surprising that the truncation of the last three residues of the CD1a cytoplasmic tail did not affect trafficking, including the rate of internalization and recycling. This suggests that the pathway followed by CD1a does not require any specific motifs in the cytoplasmic tail. In the absence of specific motifs to direct intracellular trafficking, the pathway followed by CD1a could represent what might be described as a default recycling pathway for cell surface proteins. Alternatively, CD1a could interact with an unknown protein other than Ii [Sloma et al. 2008] that regulates its trafficking along the endocytic pathway. Furthermore, we did not detect any defects in the antigen-presenting capacity of CD1a TD compared with CD1a WT, suggesting that the S-acylation of the CD1a cytoplasmic tail does not play a significant role in the function of the protein. However, CD1a has been shown to be capable of loading antigens on the cell surface [Manolova et al. 2006], which could mask changes in intracellular loading of antigens onto CD1a.

Earlier we have shown that MHC class II/Ii associates with CD1a [Sloma et al. 2008] and may influence CD1a trafficking. Our new findings complement the first study by examining directly the CD1a internalization and recycling pathway and the role of CD1a cytoplasmic tail in a cell type that does not express Ii. Furthermore, we confirmed that CD1a localizes to detergent-resistant membrane microdomains in HeLa cells, but this was not dependent on the S-acylation of its cytoplasmic tail. We also confirmed that cholesterol-depleted immature monocyte-derived DC are less efficient in presenting CD1a-dependent exogenous antigens loaded on the cell surface by using dideoxymycobactin in addition to sulfatide. This suggests that plasma membrane detergent-resistant membrane microdomains are essential for CD1a-restricted antigen presentation in DC.

We found that CD1a and MHC class I follow similar intracellular trafficking pathways. Both molecules can be detected in the Rab22a-dependent recycling pathway and recycle in an ARF6-dependent manner, contrary to TfR and other cargo internalized in a clathrin-dependent fashion. Indeed, we show in HeLa cells and DC that the internalization of CD1a is essentially clathrin, AP-2 and dynamin-independent. The reduction in CD1a surface levels, seen only in DC when the AP-2 m2 subunit was knocked down, was unexpected. One possibility is that DC might upregulate clathrin-independent pathways of internalization, such as the one followed by CD1a, when the clathrin-mediated pathway is blocked. Our study looked for the first time at the internalization of CD1a, and we were surprised to find that the kinetics were very fast reaching equilibrium after 20 min, a rate matching receptor-mediated internalization. This implies that the clathrin-independent pathway followed by CD1a mediates fast internalization, similar to the clathrin-dependent pathway.

It is generally accepted that CD1 evolved from a classical MHC class I ancestor early in vertebrate evolution [Dascher 2007]. In this context, CD1a could represent a more primordial CD1 isoform, when compared for example with CD1b, because it has a smaller antigen-binding groove and binds lipids with short alkyl chains (e.g. sulfatide) [Shamshiev et al. 2002; Zajonc et al. 2003]. It also can bind lipids on the surface of the cell, contrasting with CD1b, which can accommodate lipids with C_{80} alkyl chains in its antigen-binding groove and can only load these long lipids in acidic compartments such as LE and lysosomes [Manolova et al. 2006; Moody et al. 2002]. A model has been proposed where, after gene duplication, different CD1 isoforms acquired different trafficking patterns, which then led to changes in the antigen-binding groove, to accommodate lipids that were present in the newly surveyed intracellular compartments [Dascher & Brenner 2003]. Our results fit this model in the sense that CD1a shows a similar trafficking pattern to MHC class I. Recently, the presence of two CD1 genes in chicken was reported, which makes these the most ancient known so far [Miller et al. 2005; Salomonsen et al. 2005].

13.16. Conclusions and outlook

Interestingly, one of these proteins shows an intracellular distribution similar to human CD1a (D. C. Barral, M. B. Brenner and C. C. Dascher, unpublished data). Therefore, our study describes another level of similarity between MHC class I and CD1a, namely their intracellular trafficking, and these molecules essentially identify an internalization and recycling pathway that is only partially characterized. Yet, this pathway may be followed by a number of cell surface molecules with important immunological functions.

Our third study [Cernadas et al. 2009] investigates the pathways followed by CD1a and their intersection with lipid pathways of potential antigens. Successful antigen presentation involves an orchestrated series of steps that are critically dependent on the intersection of antigens with the relevant antigen presenting molecules. Lessons from the MHC class I and MHC class II pathways have made it clear that access to the proper antigen loading compartment is a key feature for those presentation systems to succeed. The proteolytic environment present in lysosomes plays a critical role in the ability of a cell to process protein antigens for prolonged antigen presentation. APC such as macrophages with high levels of lysosomal proteases rapidly degrade internalized proteins reducing their capacity for sustained antigen presentation compared with DC [Delamarre et al. 2005]. Although the ability of MHC class I and II molecules to survey different intracellular compartments and antigens is a hallmark of these systems, the role of CD1 trafficking in the presentation of exogenous and endogenous antigens has been examined in only a few cases.

The unique intracellular distribution of each CD1 isoform has been appreciated for some time. It has been demonstrated for the CD1b and CD1c isoforms that their intracellular trafficking is directed by their respective tyrosine-based motifs using tail-deleted constructs [Briken et al. 2000; Sugita et al. 2002]. Tail-deleted CD1b molecules have been demonstrated to have significant impairment in their antigen presenting abilities *in vitro* [Sugita et al. 2002]. Glucose monomycolate possessing a long lipid tail has been shown to require the acidic environment found in the late endocytic system to be loaded efficiently [Cheng et al. 2006; Moody et al. 2002].

CD1a is the only CD1 isoform that does not have a tyrosine-based motif and almost exclusively localizes to the early endocytic and recycling pathways. Here, we demonstrate that the localization of CD1a to the early endosomal and recycling compartments is critical for the efficient and sustained presentation of specific antigens that also predominantly localize to the same endosomal sites. CD1a and sulfatide are almost exclusively colocalized to early endocytic and recycling compartments. The redirection of CD1a to LE by the addition of the transmembrane and/or cytoplasmic CD1b tyrosine-based motif to the extracellular domain of CD1a resulted in loss of colocalization with sulfatide and a correspondingly significant reduction in the ability of these transfectants to stimulate CD1a-restricted sulfatide-specific T cells. Since the extracellular domains of the chimeric molecules are identical, the most likely explanation for these findings is the difference in trafficking imparted by the chimeric cytoplasmic tails.

The activity and localization of endogenous sulfatases suggests that even if some sulfatide is delivered to lysosomes it would be rapidly degraded. There are two classes of sulfatases, catabolic and synthetic. The catabolic sulfatases are localized in lysosomes and exert their activity at acidic pH on a wide range of molecules including sulfolipids and glycosoaminoglycans. Accumulation of these molecules occurs in the setting of lysosomal sulfatase deficiency and leads to human disease including several mucopolysaccharidoses and metachromatic leukodystrophy [Diez-Roux & Ballabio 2005]. The prolonged ability of CD1a molecules to present sulfolipids such as sulfatide may be critically dependent on their localization to early endocytic compartments where the sulfatide antigen is not degraded by sulfatases. The absence of a tyrosine-based motif allows CD1a to traffic and recycle through the early endocytic system where it can most efficiently present lipid antigens that preferentially localize there or that would be destroyed either enzymatically or as a result of acidic pH in the late endocytic system. This is supported by the enhanced ability of WT CD1a to present sulfatide compared to the CD1aab tail chimeras and the sustained ability of sulfatide-pulsed DC to activate CD1a- compared to CD1b- and CD1c-restricted sulfatide-specific T cells. Additional interactions or pathways may contribute to CD1a antigen presentation as well. For example, the colocalization of CD1a with the unique Langerhans cell-specific C-type lectin required for Birbeck granule formation, langerin, in the eponymous endosomal subcompartment has been shown to augment CD1a-mediated antigen presentation [Hunger et al. 2004]. We cannot rule out the possibility that the CD1aab-sulfatide complexes dissociate in the lysosome. We should also mention the possibility that the anti-sulfatide mAb may not be able to detect sulfatide in the lysosome secondary to changes to the antigenic epitope recognized. However, similar sulfatide colocalization patterns have been observed with other anti-sulfatide mAb making this less likely.

The differential localization of CD1a to the early endocytic compartment may be important *in vivo*. The efficient and sustained presentation of antigens such as sulfatide by CD1a$^+$ DC during their migration to regional lymph nodes is likely critical for maximizing T cell activation and proliferation. This may be even of greater importance *in vivo* where antigen concentrations are likely lower than in our studies where higher concentrations of sulfatides were added to allow for antibody detection and localization by microscopy.

The specific intracellular distribution of CD1 molecules and their antigens may also explain the evolutionary diversity of the CD1 isoforms [Dascher 2007]. It is interesting to speculate that the distribution of the different CD1 isoforms throughout the endocytic system may have developed to allow for the presentation of differentially distributed antigens or antigens that are destroyed either enzymatically or by the acidic microenvironment of the late endocytic system. The CD1a, b, c, and d isoforms may have developed to maximize the efficient and sustained antigen presentation of endogenous and exogenous lipids with differences in both intracellular distribution and half-life.

These studies also highlight the importance of the differential intracellular trafficking of lipid antigens in immunity. This was demonstrated by the specific intracellular localization of sulfatide to early endocytic and recycling pathways and the receptor-mediated internalization of LAM to LE. The biophysical properties of lipid antigens, both the headgroup and acyl chains have been demonstrated to be important in antigen presentation. Interaction between the glycan headgroups of CD1a- and CD1b-restricted antigens such as *M. leprae* lipid antigens and LAM and the C-type lectins langerin and mannose receptor, respectively, have been demonstrated to play important roles in antigen uptake and presentation [Hunger et al. 2004; Prigozy et al. 1997]. The length and saturation of the acyl chain of the glycosphingolipid glucose monomycolate and GM1 length have been demonstrated to be important in CD1b-mediated antigen presentation [Cheng et al. 2006; Shamshiev et al. 2000]. Lipid length, saturation and headgroups have also been demonstrated to contribute to the endocytic sorting of lipids and the preferential retention of certain lipids in organelles [Holthuis & Levine 2005; Mukherjee & Maxfield 2000].

CD1-mediated sulfatide antigen presentation may be of increased importance in APC that express CD1a and in the setting of increased sulfatide availability. Upregulation of the production of sulfatide and other glycosphingolipids in activated DC has already been shown to play a role in CD1-mediated immunity [De Libero et al. 2005 *a*]. The high levels of expression of CD1a, long a marker of Langerhans cells, on myeloid DC point to its importance in self and foreign lipid antigen presentation. Further, since so many pathways of internalization of exogenous antigens traverse the early endocytic system, the role of CD1a as the principal lipid antigen presenting molecule in this location highlights its distinctive role as the main sentry at this entry point.

14. The MRP5 transporter is required for transport of TCR $\gamma\delta$ non-peptidic ligands

T lymphocytes expressing the T cell receptor (TCR) $\gamma\delta$ provide immediate protection and proinflammatory response during the first hours of infections. The vast majority of human $\gamma\delta$ T cells is activated by isopentenyl pyrophosphate (IPP) and other phosphorylated metabolites, generated in the cytosol of a cell. How these antigens interact with the TCR $\gamma\delta$ and a potentially involved antigen-presenting molecule (APM) remains unknown.

We found that the multidrug resistance-associated protein 5 (MRP5) transporter encoded by the *ABCC5* gene is involved in presentation of phosphoantigens, as indicated by the inhibitory activity on antigen presentation of drugs blocking MRP5 as well as by *ABCC5* gene inactivation and overexpression. MRP5 transports IPP within the endoplasmic reticulum (ER) lumen and early endosomes (EE) as suggested by its accumulation in the membranes of these compartments, thus optimizing IPP secretion. This mechanism of action resembles that of transporter associated with antigen processing (TAP) peptide transporter and depicts a similar strategy used for antigen presentation to TCR $\alpha\beta$ and $\gamma\delta$ cells.

T cell receptor (TCR) $\gamma\delta$ cells are involved in the first line of defense during infections, in antitumor immunity, in control of allergic reactions and in autoimmune diseases. These immunological functions are dependent on the release of a large variety of soluble factors, which facilitate or inhibit the immune response, and on their capacity to kill target cells. The vast majority of circulating human TCR $\gamma\delta$ cells express Vγ9 (also called Vγ2) and Vδ2 chains, characterized by junctional sequences in which N-nucleotides and recurrent J gene segments are frequently used. T cells expressing the Vγ9/Vδ2 TCR heterodimer are rare in thymus and placental blood [De Libero et al. 1991; Groh et al. 1989]. They increase over time after birth and may represent up to 10% of total circulating T cells in adults [Parker et al. 1990]. The expansion in the periphery has been ascribed to continuous stimulation by exogenous microbial and endogenous self ligands during an individual's lifetime [De Libero 1997]. Indeed, this population of TCR $\gamma\delta$ cells, which is present in primates but not in rodents, is activated by a family of self and microbial non-peptidic ligands, that are prenylated and pyrophosphorylated intermediate metabolites of the mevalonate and 2-C-methylerythritol 4-phosphate (MEP) pathways in eukaryotic and microbial cells, respectively. The most active self-ligand known is isopentenyl pyrophosphate (IPP) [Buerk et al. 1995; Gober et al. 2003], which is an important intermediate generated during cholesterol synthesis, whereas the most stimulatory microbial ligand is (E)-4-hydroxy-3-methyl-but-2-enyl pyrophosphate (HMBPP) [Hintz et al. 2001], which is an intermediate of the MEP pathway.

Because of these antigenic specificities, TCR Vγ9/Vδ2 cells are activated by tumor cells accumulating high levels of IPP [Gober et al. 2003], by cells infected with microbes, that accumulate IPP as consequence of mevalonate pathway dysregulation [Kistowska et al. 2008], as well as by exogenous HMBPP [Hintz et al. 2001], released by dying microbial cells. With the exception of the latter case, when the stimulatory ligands are freed into the extracellular space, in tumor cells or in infected cells, self antigens are generated within the cytoplasm. The site of ligand generation is the cytosolic leaflet of the endoplasmic reticulum (ER), where the enzymes of the mevalonate pathway are localized [van Meer et al. 2008]. The mechanism how phosphorylated metabolites are recognized by the TCR $\gamma\delta$ has not been evidenced so far. Stimulation with both IPP and HMBPP occurs only upon contact with other cells [De Libero et al. 1991; Lang et al. 1995], that bind these compounds [Sarikonda et al. 2008] and behave as APC. The nature of the putative antigen-presenting molecule (APM), albeit not identified to date, is indirectly indicated by its exerted function as a ubiquitous, non-polymorphic, non-β_2m-associated and species-specific molecule [De Libero et al. 1991; Gober et al. 2003; Sarikonda et al. 2008; Wei et al. 2008].

Antigenicity of phosphoantigens depends on two important features. First, the two phosphates present in IPP and HMBPP structures are essential for TCR $\gamma\delta$ cell stimulation and synthetic analogues with three, one, or no phosphates behave as weak agonists or are not stimulatory at all [Tanaka et al. 1995]. Secondly, IPP is synthesized within the cytoplasm and its negative charges pose severe limitations to free membrane diffusion. Both of these arguments raise the important question of how IPP reaches the cell surface of an APC, where it associates with the putative APM. Because of its negative charges it is very difficult to imagine passive transport of IPP through the cell membrane despite its low molecular mass. Instead, it is more probable that IPP molecules are translocated by an active inside-out transport mechanism. It is rather unlikely that the transport of endogenous IPP occurs via exocytosis from late endosomes (LE)/lysosomes due to the fact that the low pH of these endosomal compartments is detrimental for the pyrophosphate group [Van Wazer 1964] and thereby IPP immunogenicity is lost [Morita et al. 2001; Tanaka et al. 1995]. Hence, we considered the hypothesis that the membrane transport of phosphorylated antigenic metabolites is mediated by a transporter protein located within cell membranes.

14.1. The transporter of TCR $\gamma\delta$ ligands belongs to the ABC family

The type of the putative transport protein was investigated using a series of drugs inhibiting known transporters. Main objective of our investigation were three families of organic anion transporters, namely ATP-binding cassette (ABC), monocarboxylate, and inorganic transporters. In addition, we also studied the potential energy-dependence of the putative transport protein by assessing the involvement of different classes of ATPases.

All drugs, inhibiting candidate transporters, were tested for their capacity to block TCR $\gamma\delta$ cell activation. Daudi cells, used as APC, were first treated with zoledronate (ZOL) to induce accumulation of endogenous IPP [Gober et al. 2003], then extensively washed and incubated with each drug individually. Preliminary experiments were performed to carefully identify the maximal non-toxic drug doses allowed under these experimental conditions. To ascertain whether each drug was specifically acting on the TCR $\gamma\delta$ recognition of endogenous phosphorylated antigenic metabolites, several control experiments were conducted. First, drug influence on the response of the same TCR $\gamma\delta$ cells to exogenously added IPP was evaluated by using APC identically treated with the drug but not with ZOL in the presence of exogenous IPP. Secondly, drug inhibition of mitogen-induced activation of TCR $\gamma\delta$ cells was tested by using the same drug-treated APC and phytohemagglutinin mitogen, which induces T cell activation independently of TCR specificity. Thirdly, a TCR $\alpha\beta$ clone was stimulated with its specific peptide antigen in the presence of each drug. The effects of the inhibitory drugs were considered specific when TCR $\gamma\delta$ cell activation with PHA and activation of TCR $\alpha\beta$ cells were inhibited less than 20%. In case of drugs that were toxic when kept in culture for prolonged time, pulsing experiments were performed and toxic drugs were washed out before addition of T cells. The percentage of inhibition of T cell activation was calculated for each tested drug (see Table 14.1, p.139). Several drugs with a specific rather than a general effect on antigen presentation were identified, pointing to certain characteristics of a candidate transporter.

The activation of TCR $\gamma\delta$ cells was not influenced by isopentenol, geraniol, or farnesol (the alcohol analogs of stimulatory IPP) or by geranyl diphosphate or farnesyl diphosphate (data not shown); suggesting that the putative transporter selectively interacts with the negatively charged pyrophosphate group of none but IPP [Constant et al. 1994; Pfeffer et al. 1990; Schoel et al. 1994; Tanaka et al. 1994]. Furthermore, neither inhibitors of inorganic phosphate transporters (foscarnet) [Timmer & Gunn 1998; Yusufi et al. 1986] nor of monocarboxylate transporters (lactate, phloridzin and phenylpyruvate) [Halestrap & Price 1999] were able to diminish TCR $\gamma\delta$ cell activation.

On the contrary, 4,4'-diisothiocyanatostilbene-2,2'-disulfonic acid (DIDS), a potent inhibitor of anion translocators, specifically reduced ZOL-induced activation, but neither stimulation by IPP nor by PHA. These findings additionally support the hypothesis that an anion transporter is involved in the stimulation of TCR $\gamma\delta$ cells.

14.1. The TCR gamma-delta-ligand transporter belongs to the ABC family

Table 14.1 Drugs blocking transport proteins inhibit $\gamma\delta$ T cells

Drug [i]	Dose [ii]	% $\gamma\delta$ T inhibition			% $\alpha\beta$ T inhibition
		ZOL	IPP	PHA [iii]	antigen
Isopentenol	200 µM	0	0	0	nd [iv]
Geraniol	40 µM	0	0	0	nd
Farnesol	20 µM	0	0	0	nd
Foscarnet	300 mM	0	0	0	0
Lactate, Phloridzin, Phenylpyruvate	100 µM	0	0	0	nd
DIDS [v]	300 µM	40	20	15	0
ARL-67156, β,γ-MAT	300 µM	0	0	0	0
Omeprazol	30 µg/ml	0	0	0	0
Bafilomycin A1	100 nM	0	0	0	nd
Ouabain	30 µM	0	0	0	0
Oligomycin	100 µg/ml	30	20	10	0
PSC-833	100 µM	100	70	40	35
PKF-274	100 µM	100	75	45	40
Niflumic acid	100 µM	0	0	0	0
Quercetin	100 µM	30	10	0	0
Sulfinpyrazone	10 mM	95	10	0	nd
Benzbromarone	100 µM	65	40	0	nd
Probenecid	3 mM	0	0	0	0
Methotrexate	100 µM	0	0	0	0
MK-571	300 µM	50	30	10	10

[i] For the specificity of the listed drugs see http://edoc.unibas.ch/701/ Table 4
[ii] Experiments were performed by Hans-Jürgen Gober
[iii] Phytohemagglutinin
[iv] Not determined
[v] 4,4'-diisothiocyanatostilbene-2,2'-disulfonic acid

The energy dependence of such an anion transport process was evaluated by ATPase inhibitors. ARL-67156, β,γ-MAT, Omeprazol, Bafilomycin A1, and Ouabain were not blocking TCR $\gamma\delta$ responses, thus making an involvement of ecto-ATPase, H^+-ATPase, V-type ATPase and Na/K transport ATPase unlikely. Whereas a depleting agent of intracellular ATP (oligomycin) showed specific inhibition, demonstrating the requirement of energy supplied in the form of ATP for this transport mechanism.

Two inhibitors of ABC transporters, PSC-833 and PKF-274, also decreased TCR $\gamma\delta$ cell-activation, thus suggesting the possible involvement of this ATP-dependent family of transporters. Both drugs are capable of blocking various transporters of different ABC families [Boesch et al. 1991; Lum et al. 1993; Paul et al. 1996], thus forcing additional experiments to identify the responsible protein for IPP transport.

The drugs sulfinpyrazone (SPZ) and benzbromarone (BZB) were strong and notably specific inhibitors in this type of T cell activation assay. The fact that these compounds stop the action of various ABC anion transporters [Bakos et al. 2000; Evers et al. 2000] reinforced the argument that transporters involved in TCR $\gamma\delta$ activation recognize the anionic part of the stimulatory ligands. Moreover, specific blocking was observed in the presence of quercetin, an inhibitor of ABCC transporters [Walgren et al. 2000]. Based on these findings, we narrowed our studies to the subfamily C of ABC transporters (for chromosomal localization see [Dean & Allikmets 2001 a] and for tissue distribution see [Dean et al. 2001 b] of ABC transporters), which have previously been described as important anion transporters [Miyazaki et al. 2004].

14.2. Involvement of ATP-binding cassette transporter of the C subfamily in transport of TCR $\gamma\delta$ ligands

Men contain 49 documented ABC transporter genes whereof the ABCC family is composed of 12 functional members plus 1 pseudo-gene: 10 multidrug resistant proteins (MRP) (MRP1-10, MRP10 being a pseudo-gene (*ABCC13*)-expressed truncated protein), two sulfonylurea receptors and a cystic fibrosis transmembrane conductance regulator [Zhou et al. 2008]. The cystic fibrosis transmembrane conductance regulator is a chloride channel and the sulfonylurea receptors are the ATP-sensing regulatory subunits of a complex potassium channel. The human MRP/ABCC transporters except MRP9/ABCC12 are all able to transport organic anions, such as drugs conjugated to glutathione, sulphate or glucuronate. MRP1-8 are organic anion pumps differing in substrate specificity and tissue distribution. Reports on MRP9 have been controversial and it seems to be an unusual MRP in many respects [Ono et al. 2007]: it does neither recognize any known MRP substrate nor show any N-glycosylation and most importantly, despite its messenger RNA (mRNA) being readily detected in human tissue and cells, the expressed protein is not.

As phosphoantigen presentation to TCR $\gamma\delta$ cells is exerted by APC of different tissue origin [De Libero et al. 1991; Gober et al. 2003], it is conceivable that the transporters involved in antigen trafficking have broad tissue distribution. Therefore, we limited candidate transporters to those that are ubiquitously expressed. Cystic fibrosis transmembrane conductance regulator was excluded because niflumic acid, a potent inhibitor of cystic fibrosis transmembrane conductance regulator [Scott-Ward et al. 2004], did not alter TCR $\gamma\delta$ cell stimulation.

We excluded MRP7 and MRP8 as well as MRP2, MRP3, and MRP6 [Borst et al. 1999] due to their low abundance and/or limited tissue distribution and focused on MRP1, MRP4, and MRP5 transporters, that are widely expressed in the body. The leukotriene analog MK-571, that specifically inhibits active transport of MRP1, MRP4, and MRP5 [Reid et al. 2003], blocked the activation of TCR $\gamma\delta$ and not of TCR $\alpha\beta$ cells. Probenecid and methotrexate, which are two potent MRP1 inhibitors [Bakos et al. 2000], were not inhibitory, thus indicating that MRP1 is not involved in IPP transport and presentation. These results restricted our studies to MRP4 and MRP5.

MRP proteins are functionally characterized by structural features. MRP1, 2, 3, 6, and 7 possess an amino-terminal transmembrane domain and a P-glycoprotein-like core consisting of two transmembrane domains and two nucleotide-binding domains. On the other hand MRP4, 5, 8, and 9 lack the amino-terminal transmembrane domain (denoted TMD0) found in the other MRP [Tusnady et al. 2006]. MRP1 mutants deleted for the TMD0 show comparable substrate transport levels to the full length protein but display impaired trafficking to the plasma membrane, whereas MRP1 with deletions in cytoplasmic loop 3 do not show altered trafficking but differ in substrate recognition and transport activity [Frelet & Klein 2006]. The same is seen for MRP2, where the TMD0 domain is required for correct targeting to the apical membrane but is not required for transport function. Being close structural relatives, only MRP5 but not MRP4 has a ubiquitous tissue distribution [Kool et al. 1997] and even more crucially, MRP4 is absent from thymus [Borst et al. 2007]. A quantitative RT-PCR (reverse transcription polymerase chain reaction) analysis was conducted to discriminate MRP4 and MRP5 mRNA levels in cell lines of different tissue origin which stimulate TCR $\gamma\delta$ cells (see Table 14.2, p.141).

Table 14.2 MRP4 and MRP5 expression in $\gamma\delta$ T cell-stimulatory cell lines

Cell line [i]	MRP4	MRP5
A-375	positive	positive
A-431	*negative*	positive
CEM 1.3	positive	positive
Colo-201	positive	positive
Daudi	*negative*	positive
HepG2	positive	positive
HL-60	positive	positive
HuH6	positive	positive
K562	positive	positive

[i] Experiments were performed by Hans-Jürgen Gober

Importantly, MRP4 mRNA was absent in A-431 and Daudi cells which very efficiently activate TCR $\gamma\delta$ cells. In contrast, MRP5 mRNA was detected in all cell lines (primer sequences are given in http://edoc.unibas.ch/701/). In another set of experiments Daudi cells appeared negative for the expression of MRP7, 8 and 9 genes (data not shown), thus formally excluding these transporters as involved in TCR $\gamma\delta$ stimulation.

14.3. MRP5 overexpression increases stimulation of TCR $\gamma\delta$ cells

To confirm that MRP5 is the transporter of the stimulatory ligands, a MRP5-overexpressing Daudi cell line was generated. The stimulatory capacity of MRP5-transfected Daudi cells is significantly increased over non-transfected cells (see Figure 14.1 (A), p.142).

A-375 cells are strong stimulators of TCR $\gamma\delta$ cells when pulsed with ZOL. Therefore, we transfected A-375 cells with the MRP5 gene and tested their stimulatory capacity (see Figure 14.1 (B,C), p.142). MRP5 overexpression facilitates TCR $\gamma\delta$ cells activation at all ZOL doses, suggesting that a more effective transport contributes to increased presentation of endogenous antigens (see Figure 14.1 (B), p.142). The stimulatory effect of MRP5-transfected cells is limited to TCR $\gamma\delta$ cells because the presentation of MART peptide to a MART-specific and major histocompatibility complex (MHC) class I-restricted TCR $\alpha\beta$ clone is not affected (see Figure 14.1 (C), p.142). To further investigate the involvement of MRP5 in T cell activation, APC were treated with a fixed dose of ZOL to induce accumulation of endogenous ligands and incubated with increasing doses of the MRP5-inhibitor sulfinpyrazone. MRP5 transfectants require three times more sulfinpyrazone than WT cells to observe the same level of T cell inhibition, thus confirming the specificity of this drug and further supporting the involvement of MRP5 in transport of endogenous ligands (see Figure 14.2, p.143). Neither exogenous addition of antigen (IPP) nor a control mitogen (PHA, data not shown) stimulation is affected by SPZ.

14.4. MRP5 gene silencing affects stimulation of TCR $\gamma\delta$ cells

In order to further validate the involvement of MRP5 in the transport of TCR $\gamma\delta$ stimulatory ligands, we knocked down the MRP5 protein by specific gene silencing. Four different short

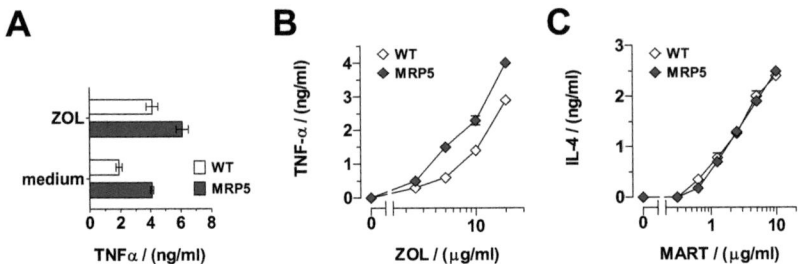

Figure 14.1
Gamma-delta T cell-activation is increased by MRP5 overexpression. (A) MRP5-transfected (closed bars) or WT Daudi cells (open bars) were used as APC in TCR $\gamma\delta$ cell-activation assays and MRP5 transfectants show a higher stimulatory capacity. Experiments were performed by Hans-Jürgen Gober. (B) MRP5 overexpression in A-375 cells facilitates activation of TCR $\gamma\delta$ cells by ZOL treated APC. MRP5-overexpressing A-375 cells (closed diamonds) stimulate $\gamma\delta$ T cells more efficiently than A-375 WT cells (open diamonds). (C) Stimulation of a MHC class I-restricted, MART peptide-specific TCR $\alpha\beta$ clone is not influenced by MRP5 overexpression. MRP5-transfected A-375 cells (closed diamonds) and A-375 WT cells (open diamonds) were used as APC. Experiments were performed by Magdalena Kistowska (B,C).

hairpin RNA (shRNA) interference constructs targeting MRP5 mRNA were generated. A-375 transfected separately with each of these constructs were used as APC in TCR $\gamma\delta$ activation experiments. All A-375 cells transfected with MRP5 shRNA interference constructs (sequences are given in http://edoc.unibas.ch/701/), stimulate TCR $\gamma\delta$ cells less efficiently than control non-transfected APC or APC transfected with a firefly luciferase shRNA construct (Ff1 [Paddison et al. 2002]). The decrease in T cell activation was observed when MRP5-silenced APC were treated with ZOL or pulsed with IPP (see Figure 14.4, p.145). There was a direct correlation of MRP5 levels after RNA interference and the stimulatory capacity, since cells transfected with the shRNA 3344 construct had the lowest of MRP5 mRNA levels and also showed the lowest stimulatory capacity (data not shown). A significant reduction of MRP5 expression in shRNA 3344 transfected cells was confirmed by confocal laser scanning microscopy (CLSM) (see Figure 14.3, p.144). The low antigen presentation capacity of MRP5 shRNA-transfected APC is TCR $\gamma\delta$-specific because presentation of MART peptide to a TCR $\alpha\beta$ MART-specific clone by the same APC is unchanged (see Figure 14.4 (C), p.145). Finally, the response of two other TCR Vγ9/Vδ2 clones was also reduced when shRNA 3344-expressing APC were used, thus confirming that the MRP5 transporter is important for the stimulation of this T cell population (data not shown).

14.5. MRP5 transports isopentenyl diphosphate across membranes

Inverted red blood cell membranes allow direct measurement of transport activity of individual substrates. They have been instrumental to confirm that cGMP is transported by different transporters, including MRP5 [Boadu & Sager 2000, 2004; Klokouzas et al. 2003; Sager 2004]. Inverted red blood cell membranes were able to transport radioactive IPP similar to cGMP (see

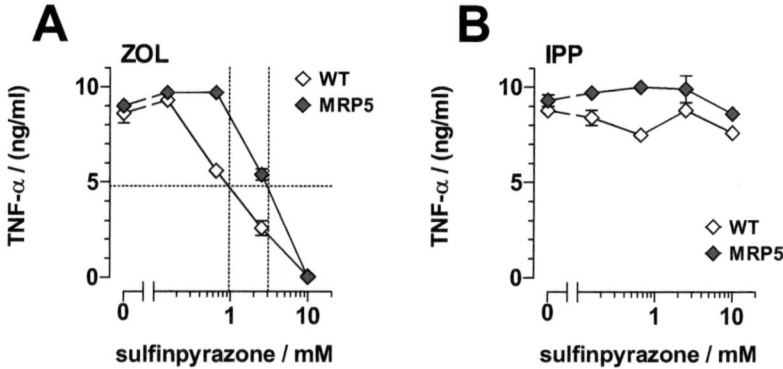

Figure 14.2
MRP5 overexpression increases resistance to ABC transporter blockers. MRP5 transfectants or WT A-375 cells were treated with ZOL (A) or pulsed with IPP (B) and used to activate TCR $\gamma\delta$ T cells in the presence of sulfinpyrazone. Exogenous addition of the TCR $\gamma\delta$-ligand IPP unalteredly stimulated $\gamma\delta$ T cells over a broad range of applied sulfinpyrazone concentrations. Conversely, ZOL treatment was less efficient on WT cells than on MRP5-overexpressing cells in the presence of sulfinpyrazone. MRP5 transfection confers resistance to sulfinpyrazone as the dose needed to reach IC_{50} levels is increased three times (x-axis difference of the two dashed vertical lines) from WT to MRP5-overexpressing cells. The dashed horizontal line represent the half-maximal stimulation. Experiments were performed by Magdalena Kistowska.

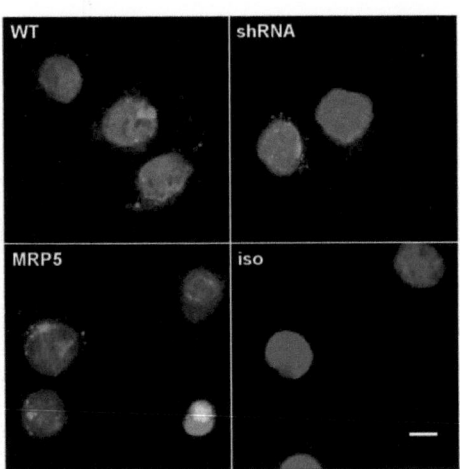

Figure 14.3
Specific silencing of MRP5 visualized by CLSM. HeLa wild type (WT, upper left), HeLa MRP5 shRNA 3344 (shRNA, upper right), and HeLa MRP5 (MRP5, lower left) were stained with polyclonal anti-human MRP5 (green) antibodies and Hoechst (red). HeLa wild type (iso, lower right) were stained with control antibodies and Hoechst. Downregulation of MRP5 expression by 3344 shRNA is obvious whereas upregulation by MRP5 transfection is hardly visible. Similar results were obtained for A-375 cells (data not shown). Scale bar, 10 µm.

14.6. MRP5 resides in ER and EE membranes

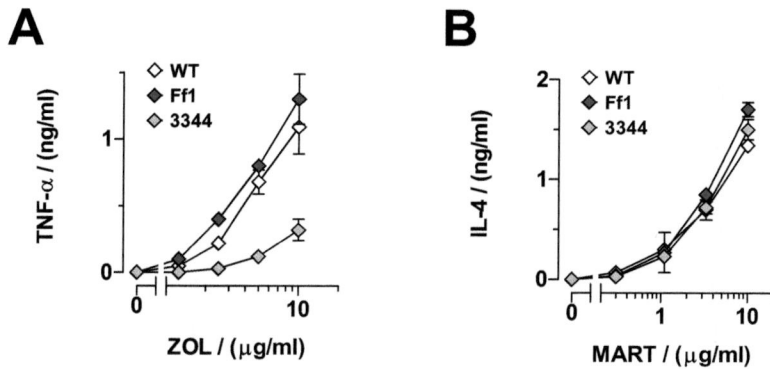

Figure 14.4
Specific silencing of MRP5 alters TCR gamma-delta cell-responses. (A and B) A-375 WT cells (white diamonds), A-375 cells transfected with the MRP5-specific shRNA interference silencing construct pCMV_3344 (light gray diamonds) or with irrelevant shRNA silencing construct Ff1 (dark gray diamonds) were used as APC in endogenous (A) antigen presentation assays. (B) The constructs did not have any impact on the activation of a MHC class I-restricted, MART peptide-specific TCR $\alpha\beta$ clone. Experiments were performed by Magdalena Kistowska.

Figure 14.5, p.146). IPP transport was inhibited by the MRP5-blocking drugs sulfinpyrazone and benzbromarone, but not by the MRP1-blocking drug methotrexate (see Figure 14.5, p.146).

IPP transport is ATP-dependent (see Figure 14.6 (left top panel), p.147). Furthermore, active transport over OSV membranes could be reduced to background levels by cold IPP (see Figure 14.6 (left bottom panel), p.147). This is likely to occur by competition for the binding site by the cold versus the hot, radioactive compound. Another possibility that we can not formally exclude is that the high amounts of IPP compete with ATP for the hydrolysis site and thus decline the energy supply for the MRP5 transporter. This would only be the case if IPP can not be hydrolyzed itself inside the ATP-binding site of MRP5. The high energy phosphate bond in pyrophosphates is a known biological short term energy depot and readily hydrolyzed under appropriate environmental conditions. Against the thought of IPP blocking the ATP-binding site is argued by the fact that the transport of cGMP, being energy- and thus ATP-dependent, is not disturbed (see Figure 14.6 (right bottom panel), p.147).

These findings confirm that MRP5 is involved in transport of IPP across membranes and raised the question of where this transport occurs within a cell other than an erythrocyte with solely a plasma membrane.

14.6. MRP5 resides in membranes of the endoplasmic reticulum and of early endosomes

Initial studies suggested that MRP5 might be localized on the plasma membrane and partly intracellularly [McAleer et al. 1999; Wijnholds et al. 1997]. Later investigations indicated that this protein is almost completely localized within the cell, although an exact localization was not

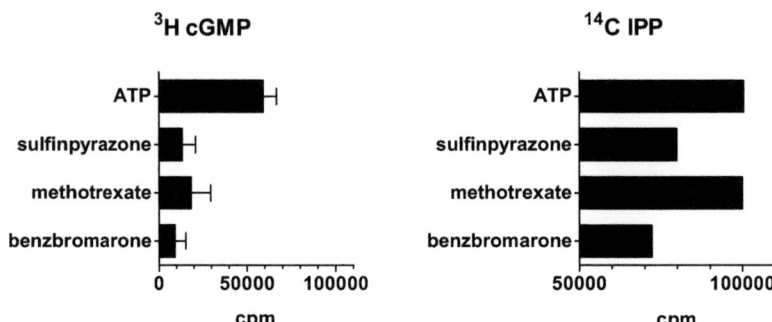

Figure 14.5
Blocking of MRP5 by specific drugs decreases IPP transport over membranes. Transport of ^{3}H cGMP control (left panel) and of ^{14}C IPP (right panel) over membranes of one-step inside-out membrane vesicles (see section 8.4 (OSV), p.49) in the presence of the ATP regenerating system (creatine phosphate, creatine phosphokinase) with ATP alone or ATP plus blocking by benzbromarone, methotrexate (both 100 µm), or sulfinpyrazone (800 µm). Benzbromarone and sulfinpyrazone inhibit transport of ^{14}C IPP and ^{3}H cGMP into inverted vesicles. Methotrexate significantly blocks transport of ^{3}H cGMP but fails to block ^{14}C IPP altogether. Uptake rates of radioactive substrates were in the range of 10 pmol/mg of protein for both compounds. SD values for ^{14}C IPP are below 400 cpm.

Figure 14.6
IPP transport by MRP5 is ATP-dependent and can be competed by cold IPP. Transport of ^{3}H IPP (left panels) and ^{3}H cGMP control (right panels) over OSV membranes in the presence of the ATP regenerating system (creatine phosphate, creatine phosphokinase) with ATP alone, with ATP-γ-S, with ATP and the inhibitory drug benzbromarone (BZB, 100 µm), or with ATP and competition by 10-fold molar excess of cold IPP. Specific, ATP-dependent transport is given by the difference of transported substrate in the presence of ATP versus ATP-γ-S (a non-hydrolyzable ATP analogue). Benzbromarone completely inhibits specific transport of ^{3}H IPP and ^{3}H cGMP into inverted vesicles. Only radioactive IPP but not cGMP is competed by cold IPP, excluding secondary effects of cold IPP to the transport system. Uptake rates of tritiated substrates were in the range of 10 pmol/mg of protein for both compounds. SD values for both substrates are less than 300 cpm. P values were lower than 0.05 for all six plots except the lower right one; cold IPP did not block ^{3}H cGMP transport.

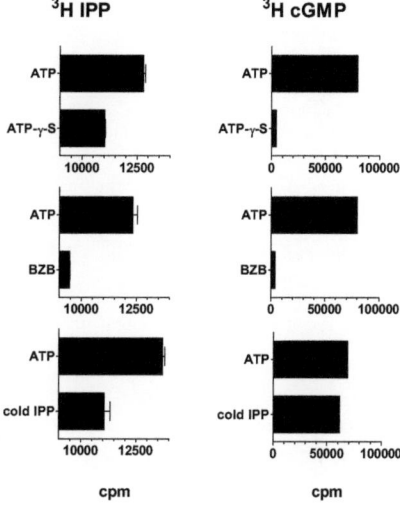

established [Torky et al. 2005]. This would be more in line with the fact that the amino-terminal transmembrane domain (TMD0) of the ABCC transporters, conferring the ability to traffic to the plasma membrane (at least for MRP1 and MRP2), is missing in MRP5.

We studied the intracellular localization of MRP5 in HeLa, A-375 and Daudi cells by confocal microscopy. In all cases, the transporter colocalized with the ER marker p63 and the transferrin receptor (TfR) in recycling early endosomes (EE) (see Figure 14.7 (row 1/2), p.149), but not with markers of the plasma membrane (data not shown), Golgi, and LE/lysosomes (see Figure 14.7 (row 3/4), p.149). To analyze the significance of the confocal studies, the Pearson's and the Mander's colocalization coefficients as well as the intensity correlation quotients of at least four complete images (with more than eight cells) per staining were calculated. Mander's M2 showed that MRP5 (green) was significantly colocalized with the TfR (EE) and p63 (ER) compartment (red) markers (see Figure 14.8, p.150). Moreover, the colocalized voxel intensities of green and red are positively (linearly) related (Person's Rtotal) and the MRP5 and compartment stainings are dependent (intensity correlation quotient).

A tubular distribution was detected, resembling that observed for other ER-localized proteins [Klopfenstein et al. 1998; Lorenz et al. 2006; Shnyrova et al. 2008; Snapp et al. 2003]. This localization shows that MRP5 is involved in antigen presentation to TCR $\gamma\delta$ cells by transporting relevant antigens within the ER and suggests that MRP5 is not directly contacting the TCR. A possible scenario is that IPP is transported to the ER lumen, where it forms complexes with a dedicated APM and, after trafficking to the cell surface, is presented to the TCR $\gamma\delta$.

14.7. Conclusions and outlook

Our data show that the immune system has hijacked a biological relevant system to transport negatively charged metabolites out of the cytoplasm. This strategy resembles that of ATP-dependent TAP involved in peptide traffic from the cytosol into ER [Momburg et al. 1994; Neefjes et al. 1993]. TAP molecules are part of complex protein multimers composed of several proteins that facilitate the assembly of MHC class I proteins and also provide transport and loading of peptides onto these presenting molecules [Cresswell et al. 1999]. At present, it is not clear whether MRP5 constitutes multi-subunit complexes with several functions including assembly of the presenting molecule, transport of phosphoantigens and formation of stimulatory antigenic complexes. The formation of such protein multimers might represent an efficient strategy to directly deliver phosphoantigens to dedicated presenting molecules.

The functional consequences arising out of MRP5 localization to EE are unclear and remain to be investigated. Lately, TAP was shown to be localized to EE and 'cross-presentation' of exogenous antigens on MHC class I molecules was demonstrated to occur in EE distinct from the ER in a TAP-dependent manner [Burgdorf et al. 2008; Gazi & Martinez-Pomares 2009].

Another non-immunological function of MRP5 might be control of the mevalonate pathway. Previous studies have identified MRP5 as a transporter for cyclic nucleotides, especially cGMP and it was suggested that MRP5 removes cGMP from the cytosol during signal transduction [Jedlitschky et al. 2000]. IPP and perhaps other phosphorylated metabolites generated in the mevalonate pathway are additional physiological substrates of MRP5. The mevalonate pathway is tightly controlled by feedback regulation of gene transcription, enzyme activity, and stability, thus accurately regulating cholesterol synthesis and protein prenylation [Brown & Goldstein 1997; Goldstein & Brown 1990; Goldstein et al. 2006]. By removing IPP accumulated in the cytosol, MRP5 might represent a novel mechanism controlling the abundance of mevalonate metabolites.

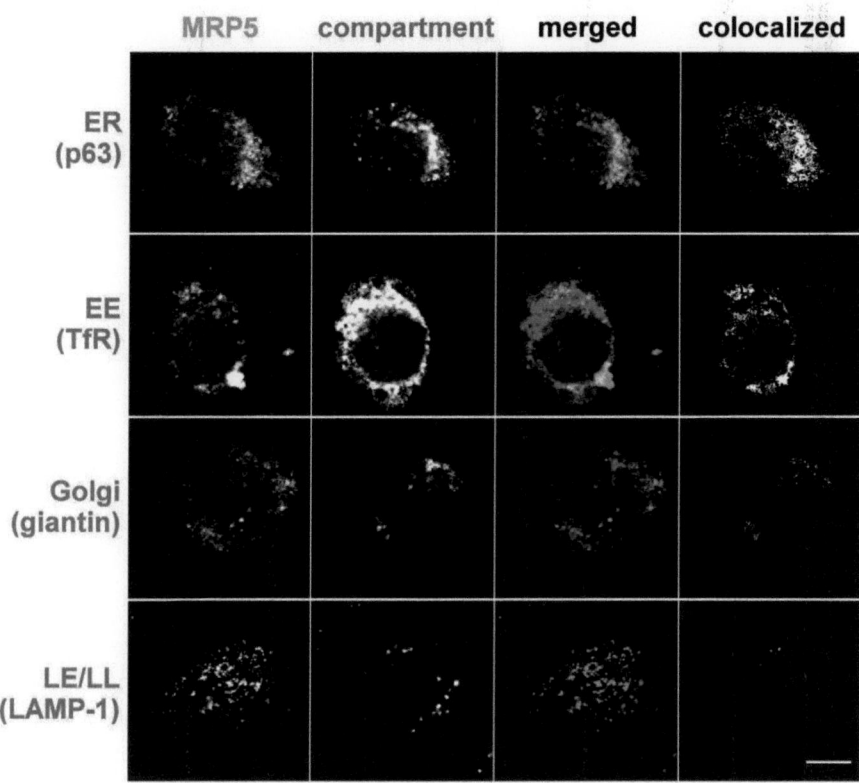

Figure 14.7
MRP5 colocalizes partially with EE and ER. Each row is showing the staining in HeLa cells of MRP5, a compartment marker, MRP5 merged to compartment all pixels, and MRP5 merged to compartment colocalized pixels only. Partial colocalization with ER (p63, first row) and with EE (TfR, second row). Neither colocalization with Golgi (giantin, third row) nor with LE/lysosomes (LAMP-1, forth row). Similar compartmentalization was seen for A-375 and Daudi cells (data not shown). Colocalized pixels were calculated in ImageJ using the Colocalization Highlighter (CH) plugin with the median channel thresholds of channel 1 and channel 2 of all images of a specific marker established by the Colocalisation Thresholds plugin and the results are shown as binary image. Scale bar, 10 µm.

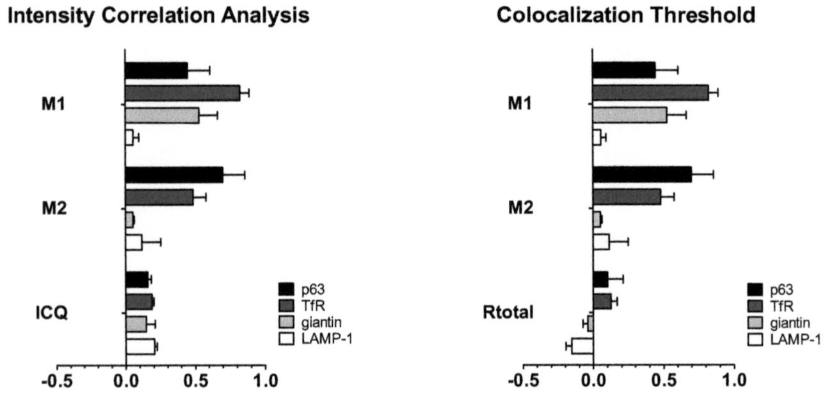

Figure 14.8
CLSM image analysis and quantification of MRP5 colocalization. Images were loaded into ImageJ, visualized using HiLo LUT for background correction (thresholding), and quantified by the Intensity Correlation Analysis and Colocalisation Thresholds plugins. At least four complete images containing more than 8 single HeLa cells without ROI selection were used to calculate Mander's Colocalization Coefficients for channel 1 (M1: e.g. red overlapping with green) and channel 2 (M2: e.g. green overlapping with red) and the Intensity Correlation Quotient (ICQ: random staining ICQ $\approx$ 0, segregated staining $0 > ICQ \geq -0.5$, dependent staining $0 < ICQ \leq +0.5$). Rtotal gives the Pearson's correlation coefficient for all the non zero-zero pixels in the image and closer it is to $\pm$ 1, the more linearly related are all the voxels' intensities in the two channels. p63 (ER), TfR (EE), giantin (Golgi), LAMP-1 (LE/lysosomes). Values are given as median $\pm$ SD.

Part IV.

General Conclusions and Future Work

It is well established that the CD1 family of proteins present lipid antigens to T cells. However, *in vivo* studies of CD1 function have been carried out almost exclusively using mouse models, despite the fact that mice lack most of the CD1 genes that are found in humans and other mammals [Dascher & Brenner 2003]. Therefore vaccination studies [Cerundolo et al. 2009; Reddy et al. 2009; Silk et al. 2004] have been hampered. Moreover lipid uptake by diet can affect the immune response to lipids [Niazi et al. 2002] but can be hardly controlled in human beings.

Considering the lymphocytes involved in non-peptidic antigen response, it would be of great worth to obtain immunological tools to monitor these cells as it has been successfully done for invariant natural killer T (iNKT) cells with lipid antigen-loaded antigen-presenting molecule (APM)-multimers. A newly established system to generate mouse iNKT cells from embryonic stem cells [Watarai et al. 2009], if applicable to human cells, will allow molecular studies about development and potentially lead to therapies with functionally selected iNKT cells.

On the antigen side, the design of new lipidic ligands was enabled by the crystal structures of the APM, thus reports on the CD1c structure as well as the identification of the TCR $\gamma\delta$-APM are eagerly awaited. Meanwhile, new non-peptidic antigens as the piperidinones or the αGC analogues with a terminal phenyl ring on the acyl chain [Schiefner et al. 2009] for the known APM will elucidate APM-TCR interactions. Optimized ligands may help to prevent immune evasion as f.i. CD1 downregulation seen by infection with *Mycobacterium tuberculosis* [Stenger et al. 1998] and to direct the immune system towards a T_H1- or T_H2-response.

Transporters for non-peptidic compounds will be more selectively studied using specific substrates as methotrexate diglutamate for MRP5 [Kruh et al. 2007] and effective translocation can be monitored by lately introduced tandem mass spectrometry methods to rapidly and sensitively detect and simultaneously quantify the eight main nonsterol intermediates of the isoprenoid biosynthesis pathway [Henneman et al. 2008]. As crystals of such multi-transmembrane transporters are unlikely to be obtained, artificial proteins [Licen et al. 2008] and molecular modeling data [Ravna et al. 2008, 2009] will be crucial to answer selectivity [Sager & Ravna 2009] and translocation of such transporters to understand their functions [Pratt et al. 2005; Warren et al. 2009]. Future MRP mouse models could be helpful and realized as mouse orthologs have been identified and studied for all human MRP genes except MRP8 [Maher et al. 2005]. The amino acid sequence of human MRP5 revealed 94.1% identity to that of mouse mrp5 and the hydrophobicity plots were almost congruent [Suzuki et al. 2000]. The *mrp5* knockout mouse has no spontaneous phenotype [Borst et al. 2007] but the $\gamma\delta$T cell compartment remains to be studied. Therefore the transgenic mouse model of the human TCR Vγ9/Vδ2 heterodimer created in our laboratory (see http://edoc.unibas.ch/701/ Part 4) might give further insights into the major human TCR $\gamma\delta$ population in PBMC normally absent in non-primates like mice.

A recent report finds MRP5 preferentially expressed in B cells and to a lower extent in T cells among the lymphocytes [Hoellein et al. 2009]. This is in line with our findings that $\gamma\delta$T cells are capable of self-stimulation when treated with ZOL or pulsed with IPP. In total PBMC the same group found MRP5 highly abundant in granulocytes (CD15- or CD16-positive cells) by quantitative real-time RT-PCR. Interestingly, eosinophils, a granulocyte subpopulation, have been proven to express functional TCR $\gamma\delta$ molecules [Legrand et al. 2009]. In allergic pleurisy a correlation of $\gamma\delta$T cells migration and activation with accumulation of eosinophils has been reported in mice [Costa et al. 2009].

Studies about lipids, their well preserved metabolic pathways, and transport systems are of great importance as they are not sole structural components of a cell but are involved in many dynamic processes as intra- and extracellular signaling [van Meer et al. 2008] as well as in immune responses by several means. They are able to strengthen an immune response by concentrating the molecules of an immune synapse locally in a membrane (raft) [Anderson et al. 2000; Eren et al. 2006; Im et al. 2009; Park et al. 2005]. Moreover, non-peptidic compounds can directly be immunogenic and presented as antigens by an antigen-presenting cell [De Libero & Mori 2009 a; Kabelitz 2008]. Furthermore, being evolutionary highly conserved, lipid metabolic pathways are ideal drug targets in pathogens as f.i. the isoprenoid biosynthetic pathways [Rohdich et al. 2005]. Overall, a longtime neglected scientific field is worth more intense efforts to clarify underlying mechanisms that will lead to new medical applications in future.

Appendices

Abbreviations

All abbreviations are designated to represent both the singular and plural forms.

Symbols

α**GC** alpha-galactosylceramide

β_2**m** β_2 microglobulin

7AAD 7-aminoactinomycin D

A

ABC ATP-binding cassette

AHA American Heart Association

AP adaptor protein

APC antigen-presenting cell

APM antigen-presenting molecule

ARF adenosine diphosphate ribosylation factor

ATH atherosclerosis

ATP adenosine triphosphate

B

BZB benzbromarone

C

CDR complementarity determining region

CFSE carboxyfluorescein succinimidyl ester

cGMP guanosine 3',5'-cyclic monophosphate

CIITA class II transactivator

CLSM confocal laser scanning microscopy

cpm counts per minute

D

DAPI 4',6-diamidino-2-phenylindole dihydrochloride

DC dendritic cell

DMAPP dimethylallyl pyrophosphate

DMSO dimethyl sulfoxide

DN double-negative

DP double-positive

E

EC endothelial cell

EC$_{50}$ half maximal effective concentration

EDTA ethylenediaminetetraacetate

EE early endosome

ER endoplasmic reticulum

F

FACS fluorescence-activated cell sorting

FCSM fetal calf serum medium

FPP farnesyl pyrophosphate

G

GFP green fluorescent protein

GM1 monosialotetrahexosylganglioside

H

HLA human leukocyte antigen

HMBPP (E)-4-hydroxy-3-methyl-but-2-enyl pyrophosphate

HSM human serum medium

I

IC$_{50}$ half maximal inhibitory concentration

IEF isoelectric focusing

Ig immunoglobulin

iGb3 isoglobotrihexosylceramide

Ii invariant chain

IL interleukin

iNKT invariant natural killer T

IPP isopentenyl pyrophosphate

IVUS intravascular ultrasound

K

KIR killer cell immunoglobulin-like receptor

KLR killer cell lectin-like receptor

L

LAMP-1 lysosome-associated membrane glycoprotein 1

LC$_{50}$ median lethal concentration

LE late endosome

M

MβCD methyl-β-cyclodextrin

mAb monoclonal antibody

MACS magnetic-activated cell sorting

MEP 2-C-methylerythritol 4-phosphate

MHC major histocompatibility complex

mRNA messenger RNA

MRP multidrug resistant protein

MW molecular weight

N

NCR natural cytotoxicity receptor

NK natural killer

P

PBMC peripheral blood mononuclear cell

PBS phosphate buffered saline

PFA paraformaldehyde

PI propidium iodide

R

RNA ribonucleic acid

rpm revolutions per minute

RT room temperature

S

SD standard deviation

shRNA short hairpin RNA

siRNA small interfering RNA

SOP standard operating protocol

SPZ sulfinpyrazone

T

T$_H$1 type 1 helper T cell

T$_H$2 type 2 helper T cell

TAP transporter associated with antigen processing

TCR T cell receptor

TD tail-deleted

Tf transferrin

TfR transferrin receptor

W

WT wild type

Z

ZOL zoledronate

Index

(E)-4-hydroxy-3-methyl-but-2-enyl pyrophosphate a natural intermediate of the 2-C-methylerythritol 4-phosphate (MEP) pathway of isopentenyl pyrophosphate biosynthesis. 26, 35, 139

αGC 20, *see* alpha-galactosylceramide

2-C-methylerythritol 4-phosphate the eponymous compound of a foreign alternative pathway of isopentenyl pyrophosphate biosynthesis. 13, 26, 139

ABC 9, *see* ATP-binding cassette

alpha-galactosylceramide the pan-iNKT cell activating glycosphingolipid isolated from *Agelas mauritianus*, an Okinawan sea sponge, in the 1990s by researchers at the Kirin Brewery in Japan and denominated KRN7000. 20, 23, 44, 47, 61, 63

ATH 14, *see* atherosclerosis

atherosclerosis hardening of an artery, typically asymptomatic for decades, specifically due to an atheromatous plaque with its three distinct components: the atheroma, cholesterol crystals and calcification; note that atherosclerosis is a form of arteriosclerosis. 14, 91

ATP-binding cassette a protein domain consisting of a catalytic (ATP hydrolyzing) core domain and a α-helical subdomain unique to ABC transporters. 9, 12, 140

benzbromarone a uricosuric therapeutic agent and potent inhibitor of CYP2C9 and MRP1-5 of the ABCC transporter family. 141

BZB 141, *see* benzbromarone

CIITA 109, *see* class II transactivator

class II transactivator a protein controlling to a large extent expression of MHC class II genes. 109

clone clones are derived from a single common ancestor cell and hence share an identical genotype. 43

CLSM 144, *see* confocal laser scanning microscopy

confocal laser scanning microscopy high resolution optical imaging with controlled depth of focus and laser scanning allowing 3D detection and reconstruction. 144

DC 18, *see* dendritic cell

dendritic cell the archetype professional antigen-presenting cell. 18, 44, 45, 107

DIMER β_2m-associated human CD1d loaded with αGC and dimerized via an anti-tag mAb. 64, 65

DN 24, *see* double-negative

double-negative neither expressing CD4 nor CD8. 24, 26, 66, 94

double-positive expressing CD4 and CD8 in unison. 18, 24, 26, 92

DP 18, *see* double-positive

EC$_{50}$ 94, *see* half maximal effective concentration

FACS 48, *see* fluorescence-activated cell sorting

fluorescence-activated cell sorting a technique for counting, examining, and sorting microscopic particles suspended in a stream of fluid. 48

GM1 25, *see* monosialotetrahexosylganglioside

half maximal effective concentration the dose of a substance required to provoke a response halfway between baseline and plateau. 94

half maximal inhibitory concentration the dose of a particular substance needed to inhibit a given biological process (or component of a process) by half. 82, 145

HLA 109, *see* human leukocyte antigen

HMBPP 26, *see* (E)-4-hydroxy-3-methyl-but-2-enyl pyrophosphate

human leukocyte antigen human leukocyte antigens were originally defined as cell surface antigens that mediate graft-versus-host disease in HLA-mismatched donors. 109

IC$_{50}$ 82, *see* half maximal inhibitory concentration

Ii 19, *see* invariant chain

IL 38, *see* interleukin

interleukin a group of cytokines first seen to be expressed by leukocytes. 38, 89, 91, 94, 99, 102–104

intravascular ultrasound a medical imaging methodology using ultrasound to visualize the endothelium of blood vessels in living individuals. 54

invariant chain a protein involved in formation, loading and deliverance of MHC class II. 19, 107, 109, 114

IPP 13, *see* isopentenyl pyrophosphate

isopentenyl pyrophosphate the eponymous compound of the self mevalonate pathway of isopentenyl pyrophosphate biosynthesis. 13, 26, 35, 137, 139–141, 144

IVUS 54, *see* intravascular ultrasound

LAMP-1 16, *see* lysosome-associated membrane glycoprotein 1

LC$_{50}$ 48, *see* median lethal concentration

lysosome-associated membrane glycoprotein 1 a marker for late endosomes (LE) and lysosomes. 16, 123, 126–128, 151, 152

MACS 40, *see* magnetic-activated cell sorting

magnetic-activated cell sorting a method for separation of various cell populations depending on their surface antigens. 40

median lethal concentration the dose of a toxic substance or radiation required to kill half the members of a tested population. 48

MEP 13, *see* 2-C-methylerythritol 4-phosphate

monosialotetrahexosylganglioside the prototype ganglioside, a member of the ganglio series of gangliosides which contain one sialic acid residue. 25, 35, 74, 112, 136

MRP 142, *see* multidrug resistant protein

multidrug resistant protein a ABBC transporter family member conferring multidrug resistance to organic anion compounds. 142

rad 1 rad is 1 cGy is 10^{-2} Gy. 46

SPZ 141, *see* sulfinpyrazone

sulfinpyrazone a uricosuric medication and potent inhibitor of MRP1-5 of the ABCC transporter family. 141, 143

T$_H$1 72, *see* type 1 helper T cell

T$_H$2 72, *see* type 2 helper T cell

type 1 helper T cell a T cell producing IFN-γ and mainly partnering with macrophages. 72, 84–87, 104, 155

type 2 helper T cell a T cell producing IL-4 and mainly partnering with B cells. 72, 76, 84–86, 104, 155

vaccination administration of antigenic material (the vaccine) to produce immunity to a disease to prevent or ameliorate the effects of infection by a pathogen. 3

variolation inoculate a person with smallpox (Variola) in a controlled manner as to minimize the severity of the infection and also to induce immunity against further infection. 3

ZOL 140, *see* zoledronate

zoledronate the synthetic aminobisphosphonate 1-hydroxy-2-imidazol-1-yl-1-phosphono-ethyl phosphonic acid blocking the active site of farnesyl pyrophosphate synthase. 140, 141, 143, 144

Bibliography

The bibliography is sorted by authors, by year and by the title of the publication.

Adams RH & Alitalo K (**2007**) "Molecular regulation of angiogenesis and lymphangiogenesis." *Nat Rev Mol Cell Biol (8:464–478)*:DOI. See p. 97.

Ades EW, Candal FJ, Swerlick RA, George VG, Summers S, Bosse DC, & Lawley TJ (**1992**) "HMEC-1: establishment of an immortalized human microvascular endothelial cell line." *J Invest Dermatol (99:683–690)*:URL. See p. 34.

Allison TJ, Winter CC, Fournie JJ, Bonneville M, & Garboczi DN (**2001**) "Structure of a human gammadelta T-cell antigen receptor." *Nature (411:820–824)*:DOI. See p. 26.

Altman JD, Moss PA, Goulder PJ, Barouch DH, McHeyzer-Williams MG, Bell JI, McMichael AJ, & Davis MM (**1996**) "Phenotypic analysis of antigen-specific T lymphocytes." *Science (274:94–96)*:URL. See p. 61.

Altman JD & Davis MM (**2003**) "MHC-peptide tetramers to visualize antigen-specific T cells." *Curr Protoc Immunol (Chapter 17:Unit 17.3)*:DOI. See p. 61.

Anderson HA, Hiltbold EM, & Roche PA (**2000**) "Concentration of MHC class II molecules in lipid rafts facilitates antigen presentation." *Nat Immunol (1:156–162)*:DOI. See pp. 131, 153.

Apostolopoulos J, Davenport P, & Tipping PG (**1996**) "Interleukin-8 production by macrophages from atheromatous plaques." *Arterioscler Thromb Vasc Biol (16:1007–1012)*:URL. See p. 103.

Appendino G, Bacchiega S, Minassi A, Cascio MG, De Petrocellis L, & Di Marzo V (**2007**) "The 1,2,3-triazole ring as a peptido- and olefinomimetic element: Discovery of click vanilloids and cannabinoids." *Angew Chem Int Ed Engl (46:9312–9315)*:DOI. See pp. 73, 74.

Arnold JN, Wormald MR, Sim RB, Rudd PM, & Dwek RA (**2007**) "The impact of glycosylation on the biological function and structure of human immunoglobulins." *Annu Rev Immunol (25:21–50)*:DOI. See p. 7.

Aslanian AM, Chapman HA, & Charo IF (**2005**) "Transient role for CD1d-restricted natural killer T cells in the formation of atherosclerotic lesions." *Arterioscler Thromb Vasc Biol (25:628–632)*:URL. See p. 89.

Bakker AH & Schumacher TNM (**2005**) "MHC multimer technology: current status and future prospects." *Curr Opin Immunol (17:428–433)*:DOI. See p. 61.

Bakos E, Evers R, Sinko E, Varadi A, Borst P, & Sarkadi B (**2000**) "Interactions of the human multidrug resistance proteins MRP1 and MRP2 with organic anions." *Mol Pharmacol (57:760–768)*:URL. See pp. 139, 140.

Barbieri L, Costantino V, Fattorusso E, Mangoni A, Aru E, Parapini S, & Taramelli D (**2004**) "Immunomodulatory alpha-galactoglycosphingolipids: Synthesis of a 2'-O-methyl-alpha-Gal-GSL and evaluation of its immunostimulating capacity." *Eur J Org Chem (468–473)*:DOI. See pp. 71, 73, 76.

Barral DC & Brenner MB (**2007**) "CD1 antigen presentation: how it works." *Nat Rev Immunol (7:929–941)*:DOI. See p. 111.

Barral DC, Cavallari M, McCormick PJ, Garg S, Magee AI, Bonifacino JS, De Libero G, & Brenner MB (**2008**) "CD1a and MHC class I follow a similar endocytic recycling pathway." *Traffic (9:1446–1457)*:DOI. See pp. 19, 111, 114, 115, 118, 119, 125, 131.

Baumruker T & Prieschl EE (**2002**) "Sphingolipids and the regulation of the immune response." *Semin Immunol (14:57–63)*:URL. See pp. 74, 81.

Beckman EM, Porcelli SA, Morita CT, Behar SM, Furlong ST, & Brenner MB (**1994**) "Recognition of a lipid antigen by CD1-restricted alpha beta+ T cells." *Nature (372:691–694)*:DOI. See pp. 3, 13.

Behar SM, Podrebarac TA, Roy CJ, Wang CR, & Brenner MB (**1999**) "Diverse TCRs recognize murine CD1." *J Immunol (162:161–167)*:URL. See p. 64.

Behar SM & Cardell S (**2000**) "Diverse CD1d-restricted T cells: diverse phenotypes, and diverse functions." *Semin Immunol (12:551–560)*:DOI. See p. 64.

Bendelac A, Savage PB, & Teyton L (**2007**) "The biology of NKT cells." *Annu Rev Immunol (25:297–336)*:URL. See pp. 24, 71.

Benlagha K, Weiss A, Beavis A, Teyton L, & Bendelac A (**2000**) "In vivo identification of glycolipid antigen-specific T cells using fluorescent CD1d tetramers." *J Exp Med (191:1895–1903)*:URL. See p. 61.

Bijlmakers MJ & Marsh M (**2003**) "The on-off story of protein palmitoylation." *Trends Cell Biol (13:32–42)*:URL. See p. 132.

Binder CJ, Chang MK, Shaw PX, Miller YI, Hartvigsen K, Dewan A, & Witztum JL (**2002**) "Innate and acquired immunity in atherogenesis." *Nat Med (8:1218–1226)*:URL. See p. 89.

Bleicher RJ & Cabot MC (**2002**) "Glucosylceramide synthase and apoptosis." *Biochim Biophys Acta Mol Cell Biol Lipids (1585:172–178)*:URL. See p. 71.

Boadu E & Sager G (**2000**) "ATPase activity and transport by a cGMP transporter in human erythrocyte ghosts and proteoliposome-reconstituted membrane extracts." *Biochim Biophys Acta (1509:467–474)*:URL. See p. 142.

– (**2004**) "Reconstitution of ATP-dependent cGMP transport into proteoliposomes by membrane proteins from human erythrocytes." *Scand J Clin Lab Invest (64:41–48)*:URL. See p. 142.

Bobryshev YV & Lord RS (**2005**) "Co-accumulation of dendritic cells and natural killer T cells within rupture-prone regions in human atherosclerotic plaques." *J Histochem Cytochem (53:781–785)*:URL. See pp. 100, 102.

Bock VD, Perciaccante R, Jansen TP, Hiemstra H, & van Maarseveen JH (**2006**) "Click chemistry as a route to cyclic tetrapeptide analogues: synthesis of cyclo-[Pro-Val-psi(triazole)-Pro-Tyr]." *Org Lett (8:919–922)*:DOI. See p. 73.

Boesch D, Gaveriaux C, Jachez B, Pourtier-Manzanedo A, Bollinger P, & Loor F (**1991**) "In vivo circumvention of P-glycoprotein-mediated multidrug resistance of tumor cells with SDZ PSC 833." *Cancer Res (51:4226–4233)*:URL. See p. 139.

Boisvert WA, Santiago R, Curtiss LK, & Terkeltaub RA (**1998**) "A leukocyte homologue of the IL-8 receptor CXCR-2 mediates the accumulation of macrophages in atherosclerotic lesions of LDL receptor-deficient mice." *J Clin Invest (101:353–363)*:URL. See p. 103.

Bonifacino JS & Traub LM (**2003**) "Signals for sorting of transmembrane proteins to endosomes and lysosomes." *Annu Rev Biochem (72:395–447)*:DOI. See pp. 19, 113.

Borg NA, Wun KS, Kjer-Nielsen L, Wilce MCJ, Pellicci DG, Koh R, Besra GS, Bharadwaj M, Godfrey DI, McCluskey J, & Rossjohn J (**2007**) "CD1d-lipid-antigen recognition by the semi-invariant NKT T-cell receptor." *Nature (448:44–49)*:DOI. See pp. 24, 69, 72, 82.

Borst P, Evers R, Kool M, & Wijnholds J (**1999**) "The multidrug resistance protein family." *Biochim Biophys Acta (1461:347–357)*:URL. See p. 140.

Borst P, de Wolf C, & van de Wetering K (**2007**) "Multidrug resistance-associated proteins 3, 4, and 5." *Pflugers Arch (453:661–673)*:URL. See pp. 140, 153.

Bradbury A, Calabi F, & Milstein C (**1990**) "Expression of CD1 in the mouse thymus." *Eur J Immunol (20:1831–1836)*:URL. See p. 15.

Brigl M & Brenner MB (**2004**) "CD1: antigen presentation and T cell function." *Annu Rev Immunol (22:817–890)*:DOI. See p. 19.

Brik A, Alexandratos J, Lin YC, Elder JH, Olson AJ, Wlodawer A, Goodsell DS, & Wong CH (**2005**) "1,2,3-triazole as a peptide surrogate in the rapid synthesis of HIV-1 protease inhibitors." *Chembiochem (6:1167–1169)*:URL. See p. 73.

Briken V, Jackman RM, Watts GF, Rogers RA, & Porcelli SA (**2000**) "Human CD1b and CD1c isoforms survey different intracellular compartments for the presentation of microbial lipid antigens." *J Exp Med (192:281–288)*:URL. See pp. 19, 131, 133.

Briken V, Jackman RM, Dasgupta S, Hoening S, & Porcelli SA (**2002**) "Intracellular trafficking pathway of newly synthesized CD1b molecules." *EMBO J (21:825–834)*:DOI. See pp. 19, 113, 118, 121.

Brodesser S, Sawatzki P, & Kolter T (**2003**) "Bioorganic chemistry of ceramide." *Eur J Org Chem (2021–2034)*:DOI. See pp. 71, 73, 76, 80.

Brodsky FM, Bodmer WF, & Parham P (**1979**) "Characterization of a monoclonal anti-beta 2-microglobulin antibody and its use in the genetic and biochemical analysis of major histocompatibility antigens." *Eur J Immunol (9:536–545)*:URL. See p. 31.

Brodsky FM, Parham P, & Bodmer WF (**1980**) "Monoclonal antibodies to HLA–DRw determinants." *Tissue Antigens (16:30–48)*:URL. See p. 32.

Brossay L, Chioda M, Burdin N, Koezuka Y, Casorati G, Dellabona P, & Kronenberg M (**1998**) "CD1d-mediated recognition of an alpha-galactosylceramide by natural killer T cells is highly conserved through mammalian evolution." *J Exp Med (188:1521–1528)*:URL. See pp. 37, 92.

Brown FD, Rozelle AL, Yin HL, Balla T, & Donaldson JG (**2001**) "Phosphatidylinositol 4,5-bisphosphate and Arf6-regulated membrane traffic." *J Cell Biol (154:1007–1017)*:DOI. See p. 119.

Brown MS & Goldstein JL (**1997**) "The SREBP pathway: regulation of cholesterol metabolism by proteolysis of a membrane-bound transcription factor." *Cell (89:331–340)*:URL. See p. 148.

Brozovic S, Nagaishi T, Yoshida M, Betz S, Salas A, Chen D, Kaser A, Glickman J, Kuo T, Little A, Morrison J, Corazza N, Kim JY, Colgan SP, Young SG, Exley M, & Blumberg RS (**2004**) "CD1d function is regulated by microsomal triglyceride transfer protein." *Nat Med (10:535–539)*:DOI. See p. 19.

Buerk MR, Mori L, & De Libero G (**1995**) "Human V gamma 9-V delta 2 cells are stimulated in a cross-reactive fashion by a variety of phosphorylated metabolites." *Eur J Immunol (25:2052–2058)*:URL. See pp. 13, 26, 137.

Bukowski JF, Morita CT, Tanaka Y, Bloom BR, Brenner MB, & Band H (**1995**) "V gamma 2V delta 2 TCR-dependent recognition of non-peptide antigens and Daudi cells analyzed by TCR gene transfer" *J Immunol (154:998–1006)*:URL. See pp. 21, 26.

Burdin N, Brossay L, Degano M, Iijima H, Gui M, Wilson IA, & Kronenberg M (**2000**) "Structural requirements for antigen presentation by mouse CD1." *Proc Natl Acad Sci U S A (97:10156–10161)*:URL. See p. 69.

Burgdorf S, Schoelz C, Kautz A, Tampe R, & Kurts C (**2008**) "Spatial and mechanistic separation of cross-presentation and endogenous antigen presentation." *Nat Immunol (9:558–566)*:DOI. See p. 148.

Calabi F & Milstein C (**1986**) "A novel family of human major histocompatibility complex-related genes not mapping to chromosome 6." *Nature (323:540–543)*:DOI. See p. 15.

– (**2000**) "The molecular biology of CD1." *Semin Immunol (12:503–509)*:DOI. See p. 15.

Canchis PW, Bhan AK, Landau SB, Yang L, Balk SP, & Blumberg RS (**1993**) "Tissue distribution of the non-polymorphic major histocompatibility complex class I-like molecule, CD1d." *Immunology (80:561–565)*:URL. See p. 18.

Cao X, Sugita M, Van Der Wel N, Lai J, Rogers RA, Peters PJ, & Brenner MB (**2002**) "CD1 molecules efficiently present antigen in immature dendritic cells and traffic independently of

MHC class II during dendritic cell maturation." *J Immunol (169:4770–4777)*:URL. See pp. 18, 107, 127.

Cardell S, Tangri S, Chan S, Kronenberg M, Benoist C, & Mathis D (**1995**) "CD1-restricted CD4+ T cells in major histocompatibility complex class II-deficient mice." *J Exp Med (182:993–1004)*:URL. See p. 23.

Carding SR & Egan PJ (**2002**) "Gammadelta T cells: functional plasticity and heterogeneity." *Nat Rev Immunol (2:336–345)*:DOI. See pp. 26, 27.

Carena I, Shamshiev A, Donda A, Colonna M, & De Libero G (**1997**) "Major histocompatibility complex class I molecules modulate activation threshold and early signaling of T cell antigen receptor-gamma/delta stimulated by nonpeptidic ligands." *J Exp Med (186:1769–1774)*:URL. See p. 27.

Carlson CM, Endrizzi BT, Wu J, Ding X, Weinreich MA, Walsh ER, Wani MA, Lingrel JB, Hogquist KA, & Jameson SC (**2006**) "Kruppel-like factor 2 regulates thymocyte and T-cell migration." *Nature (442:299–302)*:DOI. See p. 7.

Casorati G, De Libero G, Lanzavecchia A, & Migone N (**1989**) "Molecular analysis of human gamma/delta+ clones from thymus and peripheral blood." *J Exp Med (170:1521–1535)*:URL. See pp. 26, 35.

Castellino F, Galli G, Giudice GD, & Rappuoli R (**2009**) "Generating memory with vaccination." *Eur J Immunol (39:2100–2105)*:DOI. See p. 7.

Cernadas M[£], **Cavallari M**[£], Watts G, Mori L, De Libero G[$], & Brenner MB[$] (**2009**) "Early recycling compartment trafficking of CD1a is essential for its intersection and presentation of lipid antigens." *J Immunol*:DOI. See pp. 121, 125, 127, 133.

Cerundolo V, Silk JD, Masri SH, & Salio M (**2009**) "Harnessing invariant NKT cells in vaccination strategies." *Nat Rev Immunol (9:28–38)*:DOI. See p. 153.

Chan WL, Pejnovic N, Hamilton H, Liew TV, Popadic D, Poggi A, & Khan SM (**2005**) "Atherosclerotic abdominal aortic aneurysm and the interaction between autologous human plaque-derived vascular smooth muscle cells, type 1 NKT, and helper T cells." *Circ Res (96:675–683)*:URL. See p. 102.

Chang YJ, Huang JR, Tsai YC, Hung JT, Wu D, Fujio M, Wong CH, & Yu AL (**2007**) "Potent immune-modulating and anticancer effects of NKT cell stimulatory glycolipids." *Proc Natl Acad Sci U S A (104:10299–10304)*:URL. See p. 102.

Chen X, Wang X, Keaton JM, Reddington F, Illarionov PA, Besra GS, & Gumperz JE (**2007**) "Distinct endosomal trafficking requirements for presentation of autoantigens and exogenous lipids by human CD1d molecules." *J Immunol (178:6181–6190)*:URL. See p. 25.

Chen YH, Wang B, Chun T, Zhao L, Cardell S, Behar SM, Brenner MB, & Wang CR (**1999**) "Expression of CD1d2 on thymocytes is not sufficient for the development of NK T cells in CD1d1-deficient mice." *J Immunol (162:4560–4566)*:URL. See p. 25.

Cheng TY, Relloso M, Van Rhijn I, Young DC, Besra GS, Briken V, Zajonc DM, Wilson IA, Porcelli S, & Moody DB (**2006**) "Role of lipid trimming and CD1 groove size in cellular antigen presentation." *EMBO J (25:2989–2999)*:DOI. See pp. 133, 134.

Chien Yh & Bonneville M (**2006**) "Gamma delta T cell receptors." *Cell Mol Life Sci (63:2089–2094)*:DOI. See p. 26.

Chiu YH, Jayawardena J, Weiss A, Lee D, Park SH, Dautry-Varsat A, & Bendelac A (**1999**) "Distinct subsets of CD1d-restricted T cells recognize self-antigens loaded in different cellular compartments." *J Exp Med (189:103–110)*:URL. See pp. 25, 113.

Chiu YH, Park SH, Benlagha K, Forestier C, Jayawardena-Wolf J, Savage PB, Teyton L, & Bendelac A (**2002**) "Multiple defects in antigen presentation and T cell development by mice expressing cytoplasmic tail-truncated CD1d." *Nat Immunol (3:55–60)*:DOI. See pp. 19, 25, 113.

Clarke MCH, Figg N, Maguire JJ, Davenport AP, Goddard M, Littlewood TD, & Bennett MR (**2006**) "Apoptosis of vascular smooth muscle cells induces features of plaque vulnerability in atherosclerosis." *Nat Med (12:1075–1080)*:URL. See p. 89.

Cohen NR, Garg S, & Brenner MB (**2009**) "Antigen Presentation by CD1 Lipids, T Cells, and NKT Cells in Microbial Immunity." *Adv Immunol (102:1–94)*:DOI. See p. 25.

Collins SJ, Gallo RC, & Gallagher RE (**1977**) "Continuous growth and differentiation of human myeloid leukaemic cells in suspension culture." *Nature (270:347–349)*:URL. See p. 34.

Constant P, Davodeau F, Peyrat MA, Poquet Y, Puzo G, Bonneville M, & Fournie JJ (**1994**) "Stimulation of human gamma delta T cells by nonpeptidic mycobacterial ligands." *Science (264:267–270)*:URL. See pp. 13, 138.

Costa MFS, Nihei J, Mengel J, Henriques MG, & Penido C (**2009**) "Requirement of L-selectin for gammadelta T lymphocyte activation and migration during allergic pleurisy: co-relation with eosinophil accumulation." *Int Immunopharmacol (9:303–312)*:DOI. See p. 153.

Costantino V, Fattorusso E, Imperatore C, & Mangoni A (**2002**) "Immunomodulating glycosphingolipids: an efficient synthesis of a 2'-deoxy-alpha-galactosyl-GSL." *Tetrahedron (58:369–375)*:DOI. See pp. 71, 73, 76.

Costes SV, Daelemans D, Cho EH, Dobbin Z, Pavlakis G, & Lockett S (**2004**) "Automatic and quantitative measurement of protein-protein colocalization in live cells." *Biophys J (86:3993–4003)*:URL. See p. 55.

Cox D, Fox L, Tian R, Bardet W, Skaley M, Mojsilovic D, Gumperz J, & Hildebrand W (**2009**) "Determination of cellular lipids bound to human CD1d molecules." *PLoS One (4:e5325)*:DOI. See pp. 20, 25, 82.

Cresswell P, Bangia N, Dick T, & Diedrich G (**1999**) "The nature of the MHC class I peptide loading complex." *Immunol Rev (172:21–28)*:URL. See p. 148.

Cuvillier O (**2002**) "Sphingosine in apoptosis signaling." *Biochim Biophys Acta (1585:153–162)*:URL. See pp. 71, 73, 76.

Danska JS, Livingstone AM, Paragas V, Ishihara T, & Fathman CG (**1990**) "The presumptive CDR3 regions of both T cell receptor alpha and beta chains determine T cell specificity for myoglobin peptides." *J Exp Med (172:27–33)*:URL. See p. 69.

Das H, Wang L, Kamath A, & Bukowski JF (**2001**) "Vgamma2Vdelta2 T-cell receptor-mediated recognition of aminobisphosphonates." *Blood (98:1616–1618)*:DOI. See p. 13.

Dascher CC (**2007**) "Evolutionary biology of CD1." *Curr Top Microbiol Immunol (314:3–26)*:URL. See pp. 132, 134.

Dascher CC & Brenner MB (**2003**) "Evolutionary constraints on CD1 structure: insights from comparative genomic analysis." *Trends Immunol (24:412–418)*:URL. See pp. 15–17, 132, 153.

Dashtsoodol N, Watarai H, Sakata S, & Taniguchi M (**2008**) "Identification of CD4(-)CD8(-) double-negative natural killer T cell precursors in the thymus." *PLoS One (3:e3688)*:DOI. See p. 24.

De Libero G (**1997**) "Sentinel function of broadly reactive human gamma delta T cells." *Immunol Today (18:22–26)*:URL. See pp. 13, 137.

De Libero G, Casorati G, Giachino C, Carbonara C, Migone N, Matzinger P, & Lanzavecchia A (**1991**) "Selection by two powerful antigens may account for the presence of the major population of human peripheral gamma/delta T cells." *J Exp Med (173:1311–1322)*:URL. See pp. 137, 140.

De Libero G, Rocci MP, Casorati G, Giachino C, Oderda G, Tavassoli K, & Migone N (**1993**) "T cell receptor heterogeneity in gamma delta T cell clones from intestinal biopsies of patients with celiac disease." *Eur J Immunol (23:499–504)*:URL. See pp. 26, 42.

De Libero G & Mori L (**2007 a**) "Structure and biology of self lipid antigens." *Curr Top Microbiol Immunol (314:51–72)*:URL. See p. 14.

De Libero G, Moran AP, Gober HJ, Rossy E, Shamshiev A, Chelnokova O, Mazorra Z, Vendetti S, Sacchi A, Prendergast MM, Sansano S, Tonevitsky A, Landmann R, & Mori L (**2005 a**) "Bacterial infections promote T cell recognition of self-glycolipids." *Immunity (22:763–772)*:DOI. See pp. 125, 134.
De Libero G & Mori L (**2005 b**) "Recognition of lipid antigens by T cells." *Nat Rev Immunol (5:485–496)*:DOI. See p. 20.
– (**2006**) "Mechanisms of lipid-antigen generation and presentation to T cells." *Trends Immunol (27:485–492)*:DOI. See p. 9.
De Libero G, MacDonald RH, & Dellabona P (**2007 b**) "T cell recognition of lipids: quo vadis?" *Nat Immunol (8:223–227)*:DOI. See p. ii.
De Libero G & Mori L (**2009 a**) "How the immune system detects lipid antigens." *Prog Lipid Res*:DOI. See pp. 4, 9, 25, 154.
De Libero G, Collmann A, & Mori L (**2009 b**) "The cellular and biochemical rules of lipid antigen presentation." *Eur J Immunol (39:2648–2656)*:DOI. See p. 25.
de la Salle H, Mariotti S, Angenieux C, Gilleron M, Garcia-Alles LF, Malm D, Berg T, Paoletti S, Maitre B, Mourey L, Salamero J, Cazenave JP, Hanau D, Mori L, Puzo G, & De Libero G (**2005**) "Assistance of microbial glycolipid antigen processing by CD1e." *Science (310:1321–1324)*:DOI. See p. 20.
Dean M & Allikmets R (**2001 a**) "Complete characterization of the human ABC gene family." *J Bioenerg Biomembr (33:475–479)*:URL. See p. 140.
Dean M, Rzhetsky A, & Allikmets R (**2001 b**) "The human ATP-binding cassette (ABC) transporter superfamily." *Genome Res (11:1156–1166)*:URL. See p. 140.
Delamarre L, Pack M, Chang H, Mellman I, & Trombetta ES (**2005**) "Differential lysosomal proteolysis in antigen-presenting cells determines antigen fate." *Science (307:1630–1634)*:DOI. See p. 133.
Delgado A, Casas J, Llebaria A, Abad JL, & Fabrias G (**2006**) "Inhibitors of sphingolipid metabolism enzymes." *Biochim Biophys Acta (1758:1957–1977)*:URL. See pp. 73, 76.
Delia D, Cattoretti G, Polli N, Fontanella E, Aiello A, Giardini R, Rilke F, & Della Porta G (**1988**) "CD1c but neither CD1a nor CD1b molecules are expressed on normal, activated, and malignant human B cells: identification of a new B-cell subset." *Blood (72:241–247)*:URL. See p. 131.
Dellabona P, Padovan E, Casorati G, Brockhaus M, & Lanzavecchia A (**1994**) "An invariant V alpha 24-J alpha Q/V beta 11 T cell receptor is expressed in all individuals by clonally expanded CD4-8- T cells." *J Exp Med (180:1171–1176)*:URL. See p. 33.
Dere RT & Zhu X (**2008**) "The first synthesis of a thioglycoside analogue of the immunostimulant KRN7000." *Org Lett (10:4641–4644)*:DOI. See p. 71.
Diez-Roux G & Ballabio A (**2005**) "Sulfatases and human disease." *Annu Rev Genomics Hum Genet (6:355–379)*:DOI. See pp. 127, 133.
Dougan SK, Kaser A, & Blumberg RS (**2007**) "CD1 expression on antigen-presenting cells." *Curr Top Microbiol Immunol (314:113–141)*:URL. See p. 18.
Dougan SK, Salas A, Rava P, Agyemang A, Kaser A, Morrison J, Khurana A, Kronenberg M, Johnson C, Exley M, Hussain MM, & Blumberg RS (**2005**) "Microsomal triglyceride transfer protein lipidation and control of CD1d on antigen-presenting cells." *J Exp Med (202:529–539)*:DOI. See p. 19.
Doyle B & Caplice N (**2007**) "Plaque neovascularization and antiangiogenic therapy for atherosclerosis." *J Am Coll Cardiol (49:2073–2080)*:URL. See pp. 89, 97.
Driguez H (**2001**) "Thiooligosaccharides as tools for structural biology." *Chembiochem (2:311–318)*:URL. See p. 71.

Dugast M, Toussaint H, Dousset C, & Benaroch P (**2005**) "AP2 clathrin adaptor complex, but not AP1, controls the access of the major histocompatibility complex (MHC) class II to endosomes." *J Biol Chem (280:19656–19664)*:DOI. See p. 131.

Dunford JE, Thompson K, Coxon FP, Luckman SP, Hahn FM, Poulter CD, Ebetino FH, & Rogers MJ (**2001**) "Structure-activity relationships for inhibition of farnesyl diphosphate synthase in vitro and inhibition of bone resorption in vivo by nitrogen-containing bisphosphonates." *J Pharmacol Exp Ther (296:235–242)*:URL. See p. 13.

Durante-Mangoni E, Wang R, Shaulov A, He Q, Nasser I, Afdhal N, Koziel MJ, & Exley MA (**2004**) "Hepatic CD1d expression in hepatitis C virus infection and recognition by resident proinflammatory CD1d-reactive T cells." *J Immunol (173:2159–2166)*:URL. See p. 18.

Eberl G, Lees R, Smiley ST, Taniguchi M, Grusby MJ, & MacDonald HR (**1999**) "Tissue-specific segregation of CD1d-dependent and CD1d-independent NK T cells." *J Immunol (162:6410–6419)*:URL. See p. 25.

Eberl M, Hintz M, Reichenberg A, Kollas AK, Wiesner J, & Jomaa H (**2003**) "Microbial isoprenoid biosynthesis and human gammadelta T cell activation." *FEBS Lett (544:4–10)*:URL. See pp. 13, 26.

Engering A & Pieters J (**2001**) "Association of distinct tetraspanins with MHC class II molecules at different subcellular locations in human immature dendritic cells." *Int Immunol (13:127–134)*:URL. See p. 108.

Eren E, Yates J, Cwynarski K, Preston S, Dong R, Germain C, Lechler R, Huby R, Ritter M, & Lombardi G (**2006**) "Location of major histocompatibility complex class II molecules in rafts on dendritic cells enhances the efficiency of T-cell activation and proliferation." *Scand J Immunol (63:7–16)*:DOI. See pp. 129, 153.

Evers R, de Haas M, Sparidans R, Beijnen J, Wielinga PR, Lankelma J, & Borst P (**2000**) "Vinblastine and sulfinpyrazone export by the multidrug resistance protein MRP2 is associated with glutathione export." *Br J Cancer (83:375–383)*:DOI. See p. 139.

Exley MA, Hou R, Shaulov A, Tonti E, Dellabona P, Casorati G, Akbari O, Akman HO, Greenfield EA, Gumperz JE, Boyson JE, Balk SP, & Wilson SB (**2008**) "Selective activation, expansion, and monitoring of human iNKT cells with a monoclonal antibody specific for the TCR alpha-chain CDR3 loop." *Eur J Immunol (38:1756–1766)*:URL. See p. 90.

Exley M, Garcia J, Wilson SB, Spada F, Gerdes D, Tahir SM, Patton KT, Blumberg RS, Porcelli S, Chott A, & Balk SP (**2000**) "CD1d structure and regulation on human thymocytes, peripheral blood T cells, B cells and monocytes." *Immunology (100:37–47)*:URL. See p. 18.

Finn MG, Kolb HC, Fokin VV, & Sharpless KB (**2008**) "Click chemistry - Definition and aims." *Prog Chem (20:1–5)*:URL. See p. 73.

Fleiner M, Kummer M, Mirlacher M, Sauter G, Cathomas G, Krapf R, & Biedermann BC (**2004**) "Arterial neovascularization and inflammation in vulnerable patients: early and late signs of symptomatic atherosclerosis." *Circulation (110:2843–2850)*:URL. See pp. 89, 90, 100.

Fox LM, Cox DG, Lockridge JL, Wang X, Chen X, Scharf L, Trott DL, Ndonye RM, Veerapen N, Besra GS, Howell AR, Cook ME, Adams EJ, Hildebrand WH, & Gumperz JE (**2009**) "Recognition of lyso-phospholipids by human natural killer T lymphocytes." *PLoS Biol (7:e1000228)*:DOI. See pp. 25, 82.

Franchini L, Matto P, Ronchetti F, Panza L, Barbieri L, Costantino V, Mangoni A, Cavallari M, Mori L, & De Libero G (**2007**) "Synthesis and evaluation of human T cell stimulating activity of an alpha-sulfatide analogue." *Bioorg Med Chem (15:5529–5536)*:DOI. See pp. 25, 35, 36, 41.

Franck RW & Tsuji M (**2006**) "Alpha-c-galactosylceramides: synthesis and immunology." *Acc Chem Res (39:692–701)*:DOI. See p. 71.

Fredman P, Mansson JE, Rynmark BM, Josefsen K, Ekblond A, Halldner L, Osterbye T, Horn T, & Buschard K (**2000**) "The glycosphingolipid sulfatide in the islets of Langerhans in rat pancreas is processed through recycling: possible involvement in insulin trafficking." *Glycobiology (10:39–50)*:URL. See p. 125.

Frelet A & Klein M (**2006**) "Insight in eukaryotic ABC transporter function by mutation analysis." *FEBS Lett (580:1064–1084)*:DOI. See p. 140.

Gadola SD, Zaccai NR, Harlos K, Shepherd D, Castro-Palomino JC, Ritter G, Schmidt RR, Jones EY, & Cerundolo V (**2002 a**) "Structure of human CD1b with bound ligands at 2.3 A, a maze for alkyl chains." *Nat Immunol (3:721–726)*:DOI. See p. 18.

Gadola SD, Dulphy N, Salio M, & Cerundolo V (**2002 b**) "Valpha24-JalphaQ-independent, CD1d-restricted recognition of alpha-galactosylceramide by human CD4(+) and CD8alphabeta(+) T lymphocytes." *J Immunol (168:5514–5520)*:URL. See p. 25.

Gadola SD, Silk JD, Jeans A, Illarionov PA, Salio M, Besra GS, Dwek R, Butters TD, Platt FM, & Cerundolo V (**2006 a**) "Impaired selection of invariant natural killer T cells in diverse mouse models of glycosphingolipid lysosomal storage diseases." *J Exp Med (203:2293–2303)*:DOI. See p. 20.

Gadola SD, Koch M, Marles-Wright J, Lissin NM, Shepherd D, Matulis G, Harlos K, Villiger PM, Stuart DI, Jakobsen BK, Cerundolo V, & Jones EY (**2006 b**) "Structure and binding kinetics of three different human CD1d-alpha-galactosylceramide-specific T cell receptors." *J Exp Med (203:699–710)*:DOI. See pp. 26, 68.

Gajewski M, Seaver B, & Esslinger CS (**2007**) "Design, synthesis, and biological activity of novel triazole amino acids used to probe binding interactions between ligand and neutral amino acid transport protein SN1." *Bioorg Med Chem Lett (17:4163–4166)*:URL. See p. 74.

Gallati H, Pracht I, Schmidt J, Haering P, & Garotta G (**1987**) "A simple, rapid and large capacity ELISA for biologically active native and recombinant human IFN gamma." *J Biol Regul Homeost Agents (1:109–118)*:URL. See p. 33.

Gapin L, Matsuda JL, Surh CD, & Kronenberg M (**2001**) "NKT cells derive from double-positive thymocytes that are positively selected by CD1d." *Nat Immunol (2:971–978)*:DOI. See p. 24.

Gapin L (**2008**) "The making of NKT cells." *Nat Immunol (9:1009–1011)*:DOI. See p. 24.

Garcia-Alles LF, Versluis K, Maveyraud L, Tesouro Vallina A, Sansano S, Bello NF, Gober HJ, Guillet V, de la Salle H, Puzo G, Mori L, Heck AJR, De Libero G, & Mourey L (**2006**) "Endogenous phosphatidylcholine and a long spacer ligand stabilize the lipid-binding groove of CD1b." *EMBO J (25:3684–3692)*:DOI. See pp. 18, 20, 37, 38.

Gazi U & Martinez-Pomares L (**2009**) "Influence of the mannose receptor in host immune responses." *Immunobiology*:DOI. See p. 148.

Gerszten RE, Garcia-Zepeda EA, Lim YC, Yoshida M, Ding HA, Gimbrone MA, Luster AD, Luscinskas FW, & Rosenzweig A (**1999**) "MCP-1 and IL-8 trigger firm adhesion of monocytes to vascular endothelium under flow conditions." *Nature (398:718–723)*:URL. See p. 103.

Giabbai B, Sidobre S, Crispin MDM, Sanchez-Ruiz Y, Bachi A, Kronenberg M, Wilson IA, & Degano M (**2005**) "Crystal structure of mouse CD1d bound to the self ligand phosphatidylcholine: a molecular basis for NKT cell activation." *J Immunol (175:977–984)*:URL. See pp. 20, 25.

Giard DJ, Aaronson SA, Todaro GJ, Arnstein P, Kersey JH, Dosik H, & Parks WP (**1973**) "In vitro cultivation of human tumors: establishment of cell lines derived from a series of solid tumors." *J Natl Cancer Inst (51:1417–1423)*:URL. See p. 34.

Gilleron M, Stenger S, Mazorra Z, Wittke F, Mariotti S, Bohmer G, Prandi J, Mori L, Puzo G, & De Libero G (**2004**) "Diacylated sulfoglycolipids are novel mycobacterial antigens stimulating CD1-restricted T cells during infection with Mycobacterium tuberculosis." *J Exp Med (199:649–659)*:URL. See p. 14.

Gillis S & Smith KA (**1977**) "Long term culture of tumour-specific cytotoxic T cells." *Nature (268:154–156)*:URL. See p. 35.

Giroux M & Denis F (**2005**) "CD1d-unrestricted human NKT cells release chemokines upon Fas engagement." *Blood (105:703–710)*:DOI. See p. 25.

Gober HJ, Kistowska M, Angman L, Jenoe P, Mori L, & De Libero G (**2003**) "Human T cell receptor gammadelta cells recognize endogenous mevalonate metabolites in tumor cells." *J Exp Med (197:163–168)*:URL. See pp. 13, 27, 137, 138, 140.

Godfrey DI & Kronenberg M (**2004 a**) "Going both ways: immune regulation via CD1d-dependent NKT cells." *J Clin Invest (114:1379–1388)*:URL. See pp. 23, 25, 92.

Godfrey DI, Pellicci DG, & Smyth MJ (**2004 b**) "Immunology. The elusive NKT cell antigen–is the search over?" *Science (306:1687–1689)*:DOI. See p. 25.

Godfrey DI & Berzins SP (**2007**) "Control points in NKT-cell development." *Nat Rev Immunol (7:505–518)*:DOI. See pp. 23, 24, 71.

Gogolak P, Rethi B, Szatmari I, Lanyi A, Dezso B, Nagy L, & Rajnavolgyi E (**2007**) "Differentiation of CD1a- and CD1a+ monocyte-derived dendritic cells is biased by lipid environment and PPARgamma." *Blood (109:643–652)*:DOI. See p. 131.

Goldstein JL & Brown MS (**1990**) "Regulation of the mevalonate pathway." *Nature (343:425–430)*:DOI. See p. 148.

Goldstein JL, DeBose-Boyd RA, & Brown MS (**2006**) "Protein sensors for membrane sterols." *Cell (124:35–46)*:DOI. See p. 148.

Gregory S, Zilber MT, Choqueux C, Mooney N, Charron D, & Gelin C (**2000**) "Role of the CD1a molecule in the superantigen-induced activation of MHC class II negative human thymocytes." *Hum Immunol (61:427–437)*:URL. See p. 107.

Griffiths G (**2007**) "Cell evolution and the problem of membrane topology." *Nat Rev Mol Cell Biol (8:1018–1024)*:DOI. See p. 12.

Groh V, Porcelli S, Fabbi M, Lanier LL, Picker LJ, Anderson T, Warnke RA, Bhan AK, Strominger JL, & Brenner MB (**1989**) "Human lymphocytes bearing T cell receptor gamma/delta are phenotypically diverse and evenly distributed throughout the lymphoid system" *J Exp Med (169:1277–1294)*:URL. See p. 137.

Groh V, Steinle A, Bauer S, & Spies T (**1998**) "Recognition of stress-induced MHC molecules by intestinal epithelial gammadelta T cells." *Science (279:1737–1740)*:URL. See p. 27.

Gulbins E & Grassme H (**2002**) "Ceramide and cell death receptor clustering." *Biochim Biophys Acta Mol Cell Biol Lipids (1585:139–145)*:URL. See pp. 71, 73, 76.

Gumperz JE, Roy C, Makowska A, Lum D, Sugita M, Podrebarac T, Koezuka Y, Porcelli SA, Cardell S, Brenner MB, & Behar SM (**2000**) "Murine CD1d-restricted T cell recognition of cellular lipids." *Immunity (12:211–221)*:URL. See p. 61.

Gumperz JE, Miyake S, Yamamura T, & Brenner MB (**2002**) "Functionally distinct subsets of CD1d-restricted natural killer T cells revealed by CD1d tetramer staining." *J Exp Med (195:625–636)*:URL. See p. 61.

Guy K, Van Heyningen V, Cohen BB, Deane DL, & Steel CM (**1982**) "Differential expression and serologically distinct subpopulations of human Ia antigens detected with monoclonal antibodies to Ia alpha and beta chains." *Eur J Immunol (12:942–948)*:URL. See p. 33.

Halestrap AP & Price NT (**1999**) "The proton-linked monocarboxylate transporter (MCT) family: structure, function and regulation." *Biochem J (343 Pt 2:281–299)*:URL. See p. 138.

Hansson GK & Libby P (**2006**) "The immune response in atherosclerosis: a double-edged sword." *Nat Rev Immunol (6:508–519)*:URL. See p. 103.

Hansson GK (**2005**) "Inflammation, atherosclerosis, and coronary artery disease." *N Engl J Med (352:1685–1695)*:URL. See p. 89.

Haucke V & Di Paolo G (**2007**) "Lipids and lipid modifications in the regulation of membrane traffic." *Curr Opin Cell Biol (19:426–435)*:DOI. See p. 10.

Havran WL, Chien Yh, & Allison JP (**1991**) "Recognition of self antigens by skin-derived T cells with invariant gamma delta antigen receptors." *Science (252:1430–1432)*:URL. See p. 27.

Hayday AC (**2000**) "gamma-delta cells: a right time and a right place for a conserved third way of protection." *Annu Rev Immunol (18:975–1026)*:DOI. See p. 26.

Hayday A & Tigelaar R (**2003**) "Immunoregulation in the tissues by gammadelta T cells." *Nat Rev Immunol (3:233–242)*:DOI. See p. 27.

Hayes SM & Love PE (**2002**) "Distinct structure and signaling potential of the gamma delta TCR complex" *Immunity (16:827–838)*:URL. See p. 21.

Hayes SM & Love PE (**2007**) "A retrospective on the requirements for gammadelta T-cell development." *Immunol Rev (215:8–14)*:DOI. See p. 26.

Hemler ME (**2003**) "Tetraspanin proteins mediate cellular penetration, invasion, and fusion events and define a novel type of membrane microdomain." *Annu Rev Cell Dev Biol (19:397–422)*:DOI. See p. 129.

Henneman L, van Cruchten AG, Denis SW, Amolins MW, Placzek AT, Gibbs RA, Kulik W, & Waterham HR (**2008**) "Detection of nonsterol isoprenoids by HPLC-MS/MS." *Anal Biochem (383:18–24)*:DOI. See p. 153.

Herzner H, Reipen T, Schultz M, & Kunz H (**2000**) "Synthesis of glycopeptides containing carbohydrate and Peptide recognition motifs." *Chem Rev (100:4495–4538)*:URL. See p. 71.

Hess C, Means TK, Autissier P, Woodberry T, Altfeld M, Addo MM, Frahm N, Brander C, Walker BD, & Luster AD (**2004**) "IL-8 responsiveness defines a subset of CD8 T cells poised to kill." *Blood (104:3463–3471)*:URL. See p. 103.

Hintz M, Reichenberg A, Altincicek B, Bahr U, Gschwind RM, Kollas AK, Beck E, Wiesner J, Eberl M, & Jomaa H (**2001**) "Identification of (E)-4-hydroxy-3-methyl-but-2-enyl pyrophosphate as a major activator for human gammadelta T cells in Escherichia coli." *FEBS Lett (509:317–322)*:URL. See p. 137.

Hiromatsu K, Dascher CC, Sugita M, Gingrich-Baker C, Behar SM, LeClair KP, Brenner MB, & Porcelli SA (**2002**) "Characterization of guinea-pig group 1 CD1 proteins." *Immunology (106:159–172)*:DOI. See p. 16.

Hoellein A, Decker T, Bogner C, Oelsner M, Hauswald S, Peschel C, Keller U, & Licht T (**2009**) "Expression of multidrug resistance-associated ABC transporters in B-CLL is independent of ZAP70 status." *J Cancer Res Clin Oncol*:DOI. See p. 153.

Holthuis JCM & Levine TP (**2005**) "Lipid traffic: floppy drives and a superhighway." *Nat Rev Mol Cell Biol (6:209–220)*:DOI. See p. 134.

Horne WS, Yadav MK, Stout CD, & Ghadiri MR (**2004**) "Heterocyclic peptide backbone modifications in an alpha-helical coiled coil." *J Am Chem Soc (126:15366–15367)*:URL. See p. 74.

Horrobin DF (**1999**) "Lipid metabolism, human evolution and schizophrenia." *Prostaglandins Leukot Essent Fatty Acids (60:431–437)*:URL. See p. 9.

Huby RD, Dearman RJ, & Kimber I (**1999**) "Intracellular phosphotyrosine induction by major histocompatibility complex class II requires co-aggregation with membrane rafts." *J Biol Chem (274:22591–22596)*:URL. See p. 107.

Hunger RE, Sieling PA, Ochoa MT, Sugaya M, Burdick AE, Rea TH, Brennan PJ, Belisle JT, Blauvelt A, Porcelli SA, & Modlin RL (**2004**) "Langerhans cells utilize CD1a and langerin to efficiently present nonpeptide antigens to T cells." *J Clin Invest (113:701–708)*:DOI. See p. 134.

Im JS, Arora P, Bricard G, Molano A, Venkataswamy MM, Baine I, Jerud ES, Goldberg MF, Baena A, Yu KOA, Ndonye RM, Howell AR, Yuan W, Cresswell P, Chang YT, Illarionov PA, Besra GS, & Porcelli SA (**2009**) "Kinetics and cellular site of glycolipid loading control the outcome of natural killer T cell activation." *Immunity (30:888–898)*:DOI. See pp. 82, 153.

Jackman RM, Stenger S, Lee A, Moody DB, Rogers RA, Niazi KR, Sugita M, Modlin RL, Peters PJ, & Porcelli SA (**1998**) "The tyrosine-containing cytoplasmic tail of CD1b is essential for its efficient presentation of bacterial lipid antigens." *Immunity (8:341–351)*:URL. See pp. 19, 113.

James EA, LaFond R, Durinovic-Bello I, & Kwok W (**2009**) "Visualizing antigen specific CD4+ T cells using MHC class II tetramers." *J Vis Exp*:DOI. See p. 61.

Jameson J, Ugarte K, Chen N, Yachi P, Fuchs E, Boismenu R, & Havran WL (**2002**) "A role for skin gammadelta T cells in wound repair." *Science (296:747–749)*:DOI. See p. 27.

Jang JH, Kanoh K, Adachi K, & Shizuri Y (**2006**) "Awajanomycin, a cytotoxic gamma-lactone-delta-lactam metabolite from marine-derived Acremonium sp. AWA16-1." *J Nat Prod (69:1358–1360)*:URL. See p. 80.

Jayawardena-Wolf J, Benlagha K, Chiu YH, Mehr R, & Bendelac A (**2001**) "CD1d endosomal trafficking is independently regulated by an intrinsic CD1d-encoded tyrosine motif and by the invariant chain." *Immunity (15:897–908)*:URL. See pp. 19, 74, 129.

Jedlitschky G, Burchell B, & Keppler D (**2000**) "The multidrug resistance protein 5 functions as an ATP-dependent export pump for cyclic nucleotides." *J Biol Chem (275:30069–30074)*:DOI. See p. 148.

Jing Y, Gravenstein S, Chaganty NR, Chen N, Lyerly KH, Joyce S, & Deng Y (**2007**) "Aging is associated with a rapid decline in frequency, alterations in subset composition, and enhanced Th2 response in CD1d-restricted NKT cells from human peripheral blood." *Exp Gerontol (42:719–732)*:URL. See p. 92.

Julina R, Herzig T, Bernet B, & Vasella A (**2004**) "Enantioselective Synthesis of D-Erythro-Sphingosine and of Ceramide." *Helv Chim Acta (69:368–373)*:DOI. See p. 73.

Kabelitz D (**2008**) "Small molecules for the activation of human gammadelta T cell responses against infection." *Recent Pat Antiinfect Drug Discov (3:1–9)*:URL. See p. 154.

Kang SJ & Cresswell P (**2002 a**) "Calnexin, calreticulin, and ERp57 cooperate in disulfide bond formation in human CD1d heavy chain." *J Biol Chem (277:44838–44844)*:DOI. See p. 19.

– (**2002 b**) "Regulation of intracellular trafficking of human CD1d by association with MHC class II molecules." *EMBO J (21:1650–1660)*:DOI. See pp. 19, 129.

– (**2004**) "Saposins facilitate CD1d-restricted presentation of an exogenous lipid antigen to T cells." *Nat Immunol (5:175–181)*:DOI. See p. 20.

Karadimitris A, Gadola S, Altamirano M, Brown D, Woolfson A, Klenerman P, Chen JL, Koezuka Y, Roberts IA, Price DA, Dusheiko G, Milstein C, Fersht A, Luzzatto L, & Cerundolo V (**2001**) "Human CD1d-glycolipid tetramers generated by in vitro oxidative refolding chromatography." *Proc Natl Acad Sci U S A (98:3294–3298)*:DOI. See p. 61.

Karasuyama H, Kudo A, & Melchers F (**1990**) "The proteins encoded by the VpreB and lambda 5 pre-B cell-specific genes can associate with each other and with mu heavy chain." *J Exp Med (172:969–972)*:URL. See p. 37.

Kaser A, Hava DL, Dougan SK, Chen Z, Zeissig S, Brenner MB, & Blumberg RS (**2008**) "Microsomal triglyceride transfer protein regulates endogenous and exogenous antigen presentation by group 1 CD1 molecules." *Eur J Immunol (38:2351–2359)*:DOI. See p. 19.

Kawano T, Cui J, Koezuka Y, Toura I, Kaneko Y, Motoki K, Ueno H, Nakagawa R, Sato H, Kondo E, Koseki H, & Taniguchi M (**1997**) "CD1d-restricted and TCR-mediated activation of valpha14 NKT cells by glycosylceramides." *Science (278:1626–1629)*:URL. See p. 23.

Kawano T, Tanaka Y, Shimizu E, Kaneko Y, Kamata N, Sato H, Osada H, Sekiya S, Nakayama T, & Taniguchi M (**1999**) "A novel recognition motif of human NKT antigen receptor for a glycolipid ligand." *Int Immunol (11:881–887)*:URL. See p. 69.

Kawase M, Watanabe M, Kondo T, Yabu T, Taguchi Y, Umehara H, Uchiyama T, Mizuno K, & Okazaki T (**2002**) "Increase of ceramide in adriamycin-induced HL-60 cell apoptosis: detection by a novel anti-ceramide antibody." *Biochim Biophys Acta (1584:104–114)*:URL. See p. 71.

Kearney JF, Radbruch A, Liesegang B, & Rajewsky K (**1979**) "A new mouse myeloma cell line that has lost immunoglobulin expression but permits the construction of antibody-secreting hybrid cell lines." *J Immunol (123:1548–1550)*:URL. See p. 35.

Killisch I, Steinlein P, Roemisch K, Hollinshead R, Beug H, & Griffiths G (**1992**) "Characterization of early and late endocytic compartments of the transferrin cycle. Transferrin receptor antibody blocks erythroid differentiation by trapping the receptor in the early endosome." *J Cell Sci (103 (Pt 1):211–232)*:URL. See p. 115.

Kim S, Cho M, Lee T, Lee S, Min HY, & Lee SK (**2007**) "Design, synthesis, and preliminary biological evaluation of a novel triazole analogue of ceramide." *Bioorg Med Chem Lett (17:4584–4587)*:URL. See p. 74.

Kinjo Y, Wu D, Kim GS, Xing GW, Poles MA, Ho DD, Tsuji M, Kawahara K, Wong CH, & Kronenberg M (**2005**) "Recognition of bacterial glycosphingolipids by natural killer T cells." *Nature (434:520–525)*:URL. See p. 14.

Kinjo Y, Tupin E, Wu D, Fujio M, Garcia-Navarro R, Benhnia MREI, Zajonc DM, Ben-Menachem G, Ainge GD, Painter GF, Khurana A, Hoebe K, Behar SM, Beutler B, Wilson IA, Tsuji M, Sellati TJ, Wong CH, & Kronenberg M (**2006**) "Natural killer T cells recognize diacylglycerol antigens from pathogenic bacteria." *Nat Immunol (7:978–986)*:DOI. See p. 14.

Kistowska M, Rossy E, Sansano S, Gober HJ, Landmann R, Mori L, & De Libero G (**2008**) "Dysregulation of the host mevalonate pathway during early bacterial infection activates human TCR gamma delta cells." *Eur J Immunol (38:2200–2209)*:DOI. See pp. 27, 137.

Klokouzas A, Wu CP, van Veen HW, Barrand MA, & Hladky SB (**2003**) "cGMP and glutathione-conjugate transport in human erythrocytes." *Eur J Biochem (270:3696–3708)*:URL. See p. 142.

Klopfenstein DR, Kappeler F, & Hauri HP (**1998**) "A novel direct interaction of endoplasmic reticulum with microtubules." *EMBO J (17:6168–6177)*:DOI. See p. 148.

Kobayashi E, Motoki K, Uchida T, Fukushima H, & Koezuka Y (**1995**) "KRN7000, a novel immunomodulator, and its antitumor activities." *Oncol Res (7:529–534)*:URL. See p. 36.

Koch M, Stronge VS, Shepherd D, Gadola SD, Mathew B, Ritter G, Fersht AR, Besra GS, Schmidt RR, Jones EY, & Cerundolo V (**2005**) "The crystal structure of human CD1d with and without alpha-galactosylceramide." *Nat Immunol (6:819–826)*:DOI. See pp. 18, 68, 69, 73, 74, 80, 81.

Koehler G & Milstein C (**1975**) "Continuous cultures of fused cells secreting antibody of predefined specificity." *Nature (256:495–497)*:URL. See p. 35.

Kolb HC & Sharpless KB (**2003**) "The growing impact of click chemistry on drug discovery." *Drug Discov Today (8:1128–1137)*:URL. See p. 73.

Kolodgie FD, Gold HK, Burke AP, Fowler DR, Kruth HS, Weber DK, Farb A, Guerrero LJ, Hayase M, Kutys R, Narula J, Finn AV, & Virmani R (**2003**) "Intraplaque hemorrhage and progression of coronary atheroma." *N Engl J Med (349:2316–2325)*:URL. See p. 89.

Kolter T & Sandhoff K (**2005**) "Principles of lysosomal membrane digestion: stimulation of sphingolipid degradation by sphingolipid activator proteins and anionic lysosomal lipids." *Annu Rev Cell Dev Biol (21:81–103)*:DOI. See p. 20.

– (**2009**) "Lysosomal degradation of membrane lipids." *FEBS Lett*:DOI. See pp. 9, 20.

Kool M, de Haas M, Scheffer GL, Scheper RJ, van Eijk MJ, Juijn JA, Baas F, & Borst P (**1997**) "Analysis of expression of cMOAT (MRP2), MRP3, MRP4, and MRP5, homologues of the multidrug resistance-associated protein gene (MRP1), in human cancer cell lines." *Cancer Res (57:3537–3547)*:URL. See p. 140.

Kovalovsky D, Uche OU, Eladad S, Hobbs RM, Yi W, Alonzo E, Chua K, Eidson M, Kim HJ, Im JS, Pandolfi PP, & Sant'Angelo DB (**2008**) "The BTB-zinc finger transcriptional regulator PLZF controls the development of invariant natural killer T cell effector functions." *Nat Immunol (9:1055–1064)*:DOI. See p. 24.

Kronenberg M (**2005**) "Toward an understanding of NKT cell biology: Progress and paradoxes." *Annu Rev Immunol (23:877–900)*:URL. See pp. 24, 71.

Kronenberg M & Havran WL (**2007**) "Frontline T cells: gammadelta T cells and intraepithelial lymphocytes." *Immunol Rev (215:5–7)*:DOI. See p. 27.

Kruh GD, Guo Y, Hopper-Borge E, Belinsky MG, & Chen ZS (**2007**) "ABCC10, ABCC11, and ABCC12." *Pflugers Arch (453:675–684)*:DOI. See p. 153.

Kunzmann V, Bauer E, Feurle J, Weissinger F, Tony HP, & Wilhelm M (**2000**) "Stimulation of gammadelta T cells by aminobisphosphonates and induction of antiplasma cell activity in multiple myeloma." *Blood (96:384–392)*:URL. See p. 13.

Kyriakakis E[£], **Cavallari M**[£], **Andert J**[£], Philippova M, Bochkov V, Erne P, Koella C, Mori L, Biedermann BC[$], Resink TJ[$], & De Libero G[$] (**2009**) "Invariant natural killer T cells: linking inflammation and neovascularization in human atherosclerosis." *submitted to J Immunol*. See pp. 35, 53, 90, 97.

Lampson LA & Levy R (**1980**) "Two populations of Ia-like molecules on a human B cell line." *J Immunol (125:293–299)*:URL. See pp. 32, 33.

Lang F, Peyrat MA, Constant P, Davodeau F, David-Ameline J, Poquet Y, Vie H, Fournie JJ, & Bonneville M (**1995**) "Early activation of human V gamma 9V delta 2 T cell broad cytotoxicity and TNF production by nonpeptidic mycobacterial ligands." *J Immunol (154:5986–5994)*:URL. See pp. 21, 137.

Lang GA, Maltsev SD, Besra GS, & Lang ML (**2004**) "Presentation of alpha-galactosylceramide by murine CD1d to natural killer T cells is facilitated by plasma membrane glycolipid rafts." *Immunology (112:386–396)*:DOI. See p. 107.

Langley CH, Pawar DM, & Noe EA (**2005**) "Ab initio studies of the conformations of ethynyl formate, cyano formate and related ethynyl and cyano compounds." *J Mol Struct THEOCHEM (732:99–111)*:DOI. See p. 80.

Lawton AP, Prigozy TI, Brossay L, Pei B, Khurana A, Martin D, Zhu T, Spaete K, Ozga M, Hoening S, Bakke O, & Kronenberg M (**2005**) "The mouse CD1d cytoplasmic tail mediates CD1d trafficking and antigen presentation by adaptor protein 3-dependent and -independent mechanisms." *J Immunol (174:3179–3186)*:URL. See p. 113.

Lee T, Cho M, Ko SY, Youn HJ, Baek DJ, Cho WJ, Kang CY, & Kim S (**2007**) "Synthesis and evaluation of 1,2,3-triazole containing analogues of the immunostimulant alpha-GalCer." *J Med Chem (50:585–589)*:URL. See p. 74.

Legrand F, Driss V, Woerly G, Loiseau S, Hermann E, Fournie JJ, Heliot L, Mattot V, Soncin F, Gougeon ML, Dombrowicz D, & Capron M (**2009**) "A functional gammadeltaTCR/CD3 complex distinct from gammadeltaT cells is expressed by human eosinophils." *PLoS One (4:e5926)*:DOI. See p. 153.

Leoni LM, Shih HC, Deng L, Tuey C, Walter G, Carson DA, & Cottam HB (**1998**) "Modulation of ceramide-activated protein phosphatase 2A activity by low molecular weight aromatic compounds." *Biochem Pharmacol (55:1105–1111)*:URL. See p. 80.

Li A, Dubey S, Varney ML, Dave BJ, & Singh RK (**2003**) "IL-8 directly enhanced endothelial cell survival, proliferation, and matrix metalloproteinases production and regulated angiogenesis." *J Immunol (170:3369–3376)*:URL. See p. 97.

Li L, Tang XP, Taylor KG, Dupre DB, & Yappert MC (**2002**) "Conformational characterization of ceramides by nuclear magnetic resonance spectroscopy." *Biophys J (82:2067–2080)*:URL. See p. 73.

Li Q, Lau A, Morris TJ, Guo L, Fordyce CB, & Stanley EF (**2004**) "A syntaxin 1, Galpha(o), and N-type calcium channel complex at a presynaptic nerve terminal: analysis by quantitative immunocolocalization." *J Neurosci (24:4070–4081)*:URL. See p. 55.

Libby P (**2002**) "Inflammation in atherosclerosis." *Nature (420:868–874)*:URL. See p. 89.

Licen S, De Riccardis F, Izzo I, & Tecilla P (**2008**) "Artificial anion transporters in bilayer membranes." *Curr Drug Discov Technol (5:86–97)*:URL. See p. 153.

Lijnen HR (**2003**) "Metalloproteinases in development and progression of vascular disease." *Pathophysiol Haemost Thromb (33:275–281)*:URL. See p. 89.

Lim WS, Timmins JM, Seimon TA, Sadler A, Kolodgie FD, Virmani R, & Tabas I (**2008**) "Signal transducer and activator of transcription-1 is critical for apoptosis in macrophages subjected to endoplasmic reticulum stress in vitro and in advanced atherosclerotic lesions in vivo." *Circulation (117:940–951)*:URL. See p. 89.

Lorenz H, Hailey DW, & Lippincott-Schwartz J (**2006**) "Fluorescence protease protection of GFP chimeras to reveal protein topology and subcellular localization." *Nat Methods (3:205–210)*:DOI. See p. 148.

Lozzio CB & Lozzio BB (**1975**) "Human chronic myelogenous leukemia cell-line with positive Philadelphia chromosome." *Blood (45:321–334)*:URL. See p. 34.

Lum BL, Fisher GA, Brophy NA, Yahanda AM, Adler KM, Kaubisch S, Halsey J, & Sikic BI (**1993**) "Clinical trials of modulation of multidrug resistance. Pharmacokinetic and pharmacodynamic considerations." *Cancer (72:3502–3514)*:URL. See p. 139.

Macchia M, Barontini S, Bertini S, Di Bussolo V, Fogli S, Giovannetti E, Grossi E, Minutolo F, & Danesi R (**2001**) "Design, synthesis, and characterization of the antitumor activity of novel ceramide analogues." *J Med Chem (44:3994–4000)*:URL. See p. 80.

Maceyka M, Payne SG, Milstien S, & Spiegel S (**2002**) "Sphingosine kinase, sphingosine-1-phosphate, and apoptosis." *Biochim Biophys Acta (1585:193–201)*:URL. See pp. 71, 73, 76.

Magadan JG, Barbieri MA, Mesa R, Stahl PD, & Mayorga LS (**2006**) "Rab22a regulates the sorting of transferrin to recycling endosomes." *Mol Cell Biol (26:2595–2614)*:DOI. See p. 118.

Magee AI & Parmryd I (**2003**) "Detergent-resistant membranes and the protein composition of lipid rafts." *Genome Biol (4:234)*:DOI. See p. 114.

Maher JM, Slitt AL, Cherrington NJ, Cheng X, & Klaassen CD (**2005**) "Tissue distribution and hepatic and renal ontogeny of the multidrug resistance-associated protein (Mrp) family in mice." *Drug Metab Dispos (33:947–955)*:DOI. See p. 153.

Major AS, Wilson MT, McCaleb JL, Ru Su Y, Stanic AK, Joyce S, Van Kaer L, Fazio S, & Linton MF (**2004**) "Quantitative and qualitative differences in proatherogenic NKT cells in apolipoprotein E-deficient mice." *Arterioscler Thromb Vasc Biol (24:2351–2357)*:URL. See pp. 89, 102.

Major AS, Joyce S, & Van Kaer L (**2006**) "Lipid metabolism, atherogenesis and CD1-restricted antigen presentation." *Trends Mol Med (12:270–278)*:URL. See pp. 89, 102.

Malewicz B, Valiyaveettil JT, Jacob K, Byun HS, Mattjus P, Baumann WJ, Bittman R, & Brown RE (**2005**) "The 3-hydroxy group and 4,5-trans double bond of sphingomyelin are essential for modulation of galactosylceramide transmembrane asymmetry." *Biophys J (88:2670–2680)*:URL. See p. 81.

Mallaun M, Naeher D, Daniels MA, Yachi PP, Hausmann B, Luescher IF, Gascoigne NRJ, & Palmer E (**2008**) "The T cell receptor's alpha-chain connecting peptide motif promotes close approximation of the CD8 coreceptor allowing efficient signal initiation." *J Immunol (180:8211–8221)*:URL. See p. 69.

Manolova V, Hirabayashi Y, Mori L, & De Libero G (**2003**) "CD1a and CD1b surface expression is independent from de novo synthesized glycosphingolipids." *Eur J Immunol (33:29–37)*:DOI. See p. 37.

Manolova V, Kistowska M, Paoletti S, Baltariu GM, Bausinger H, Hanau D, Mori L, & De Libero G (**2006**) "Functional CD1a is stabilized by exogenous lipids." *Eur J Immunol (36:1083–1092)*:DOI. See pp. 20, 37, 108, 131, 132.

Marrack P, Scott-Browne JP, Dai S, Gapin L, & Kappler JW (**2008**) "Evolutionarily conserved amino acids that control TCR-MHC interaction." *Annu Rev Immunol (26:171–203)*:DOI. See p. 24.

Martin LH, Calabi F, & Milstein C (**1986**) "Isolation of CD1 genes: a family of major histocompatibility complex-related differentiation antigens." *Proc Natl Acad Sci U S A (83:9154–9158)*:URL. See p. 107.

Mathew T$^£$, **Cavallari M**$^£$, Billich A, Bornancin F, Nussbaumer P, De Libero G, & Vasella A (**2009 a**) "4,5,6-Trisubstituted Piperidinones as Conformationally Restricted Ceramide Analogues: Synthesis and Evaluation as Inhibitors of Sphingosine and Ceramide Kinases and as NKT Cell-Stimulatory Antigens." *Chem Biodivers (6:1688–1715)*:DOI. See pp. 70, 79, 80.

Mathew T, Billich A, Cavallari M, Bornancin F, Nussbaumer P, De Libero G, & Vasella A (**2009 b**) "Synthesis and Evaluation of Sphingolipid Analogues: modification of the Hydroxy Group at C(1) of 7-Oxasphingosine, and of the Hydroxy Group at C(1) and the Amide Group of 7-Oxaceramides." *Chem Biodivers (6:705–724)*:DOI. See pp. 36, 70, 73, 74, 76.

Mathew T, Billaud C, Billich A, Cavallari M, Nussbaumer P, De Libero G, & Vasella A (**2009 c**) "Synthesis of 7-Aza- and 7-Thiasphingosines, and Evaluation of Their Interaction with Sphingosine Kinases and with T-Cells." *Chem Biodivers (6:725–738)*:DOI. See pp. 36, 70, 76.

Matsuda JL, Naidenko OV, Gapin L, Nakayama T, Taniguchi M, Wang CR, Koezuka Y, & Kronenberg M (**2000**) "Tracking the response of natural killer T cells to a glycolipid antigen using CD1d tetramers." *J Exp Med (192:741–754)*:URL. See p. 61.

Matsuda JL, Gapin L, Fazilleau N, Warren K, Naidenko OV, & Kronenberg M (**2001**) "Natural killer T cells reactive to a single glycolipid exhibit a highly diverse T cell receptor beta repertoire and small clone size." *Proc Natl Acad Sci U S A (98:12636–12641)*:DOI. See p. 23.

McAleer MA, Breen MA, White NL, & Matthews N (**1999**) "pABC11 (also known as MOAT-C and MRP5), a member of the ABC family of proteins, has anion transporter activity but does not confer multidrug resistance when overexpressed in human embryonic kidney 293 cells." *J Biol Chem (274:23541–23548)*:URL. See p. 145.

McCarthy C, Shepherd D, Fleire S, Stronge VS, Koch M, Illarionov PA, Bossi G, Salio M, Denkberg G, Reddington F, Tarlton A, Reddy BG, Schmidt RR, Reiter Y, Griffiths GM, van der Merwe PA, Besra GS, Jones EY, Batista FD, & Cerundolo V (**2007**) "The length of lipids bound to human CD1d molecules modulates the affinity of NKT cell TCR and the threshold of NKT cell activation." *J Exp Med (204:1131–1144)*:DOI. See pp. 68, 69.

McMichael AJ, Pilch JR, Galfre G, Mason DY, Fabre JW, & Milstein C (**1979**) "A human thymocyte antigen defined by a hybrid myeloma monoclonal antibody." *Eur J Immunol (9:205–210)*:URL. See p. 15.

Melian A, Geng YJ, Sukhova GK, Libby P, & Porcelli SA (**1999**) "CD1 expression in human atherosclerosis. A potential mechanism for T cell activation by foam cells." *Am J Pathol (155:775–786)*:URL. See pp. 89, 100.

Melian A, Watts GF, Shamshiev A, De Libero G, Clatworthy A, Vincent M, Brenner MB, Behar S, Niazi K, Modlin RL, Almo S, Ostrov D, Nathenson SG, & Porcelli SA (**2000**) "Molecular recognition of human CD1b antigen complexes: evidence for a common pattern of interaction with alpha beta TCRs." *J Immunol (165:4494–4504)*:URL. See p. 23.

Miller MM, Wang C, Parisini E, Coletta RD, Goto RM, Lee SY, Barral DC, Townes M, Roura-Mir C, Ford HL, Brenner MB, & Dascher CC (**2005**) "Characterization of two avian MHC-like genes reveals an ancient origin of the CD1 family." *Proc Natl Acad Sci U S A (102:8674–8679)*:DOI. See p. 132.

Minowada J, Onuma T, & Moore GE (**1972**) "Rosette-forming human lymphoid cell lines. I. Establishment and evidence for origin of thymus-derived lymphocytes." *J Natl Cancer Inst (49:891–895)*:URL. See p. 34.

Mitchell GF (**2008**) "Selection, memory and selective memories: T cells, B cells and Sir Mac 1968." *Immunol Cell Biol (86:26–30)*:DOI. See p. 7.

Miyagawa F, Tanaka Y, Yamashita S, & Minato N (**2001**) "Essential requirement of antigen presentation by monocyte lineage cells for the activation of primary human gamma delta T cells by aminobisphosphonate antigen" *J Immunol (166:5508–5514)*:URL. See p. 27.

Miyamoto K, Miyake S, & Yamamura T (**2001**) "A synthetic glycolipid prevents autoimmune encephalomyelitis by inducing T(H)2 bias of natural killer T cells." *Nature (413:531–534)*:DOI. See pp. 71, 72.

Miyazaki H, Sekine T, & Endou H (**2004**) "The multispecific organic anion transporter family: properties and pharmacological significance." *Trends Pharmacol Sci (25:654–662)*:DOI. See p. 140.

Modica E, Compostella F, Colombo D, Franchini L, Cavallari M, Mori L, De Libero G, Panza L, & Ronchetti F (**2006**) "Stereoselective synthesis and immunogenic activity of the C-analogue of sulfatide." *Org Lett (8:3255–3258)*:DOI. See p. 36.

Momburg F, Roelse J, Haemmerling GJ, & Neefjes JJ (**1994**) "Peptide size selection by the major histocompatibility complex-encoded peptide transporter." *J Exp Med (179:1613–1623)*:URL. See p. 148.

Montoya CJ, Pollard D, Martinson J, Kumari K, Wasserfall C, Mulder CB, Rugeles MT, Atkinson MA, Landay AL, & Wilson SB (**2007**) "Characterization of human invariant natural killer T subsets in health and disease using a novel invariant natural killer T cell-clonotypic monoclonal antibody, 6B11." *Immunology (122:1–14)*:URL. See p. 90.

Moody DB, Ulrichs T, Muehlecker W, Young DC, Gurcha SS, Grant E, Rosat JP, Brenner MB, Costello CE, Besra GS, & Porcelli SA (**2000**) "CD1c-mediated T-cell recognition of isoprenoid glycolipids in Mycobacterium tuberculosis infection." *Nature (404:884–888)*:DOI. See pp. 13, 23.

Moody DB (**2006**) "The surprising diversity of lipid antigens for CD1-restricted T cells." *Adv Immunol (89:87–139)*:DOI. See pp. 113, 131.

Moody DB, Briken V, Cheng TY, Roura-Mir C, Guy MR, Geho DH, Tykocinski ML, Besra GS, & Porcelli SA (**2002**) "Lipid length controls antigen entry into endosomal and nonendosomal pathways for CD1b presentation." *Nat Immunol (3:435–442)*:DOI. See pp. 16, 132, 133.

Moody DB & Porcelli SA (**2003**) "Intracellular pathways of CD1 antigen presentation." *Nat Rev Immunol (3:11–22)*:DOI. See pp. 16, 19, 20.

Moody DB, Young DC, Cheng TY, Rosat JP, Roura-Mir C, O'Connor PB, Zajonc DM, Walz A, Miller MJ, Levery SB, Wilson IA, Costello CE, & Brenner MB (**2004**) "T cell activation by lipopeptide antigens." *Science (303:527–531)*:DOI. See pp. 13, 111, 125.

Moody DB, Zajonc DM, & Wilson IA (**2005**) "Anatomy of CD1-lipid antigen complexes." *Nat Rev Immunol (5:387–399)*:DOI. See p. 18.

Moreno PR, Purushothaman KR, Sirol M, Levy AP, & Fuster V (**2006**) "Neovascularization in human atherosclerosis." *Circulation (113:2245-2252)*:URL. See p. 89.

Morita CT, Beckman EM, Bukowski JF, Tanaka Y, Band H, Bloom BR, Golan DE, & Brenner MB (**1995 a**) "Direct presentation of nonpeptide prenyl pyrophosphate antigens to human gamma delta T cells." *Immunity (3:495-507)*:URL. See p. 21.

Morita CT, Mariuzza RA, & Brenner MB (**2000**) "Antigen recognition by human gamma delta T cells: pattern recognition by the adaptive immune system." *Springer Semin Immunopathol (22:191-217)*:URL. See p. 27.

Morita CT, Li H, Lamphear JG, Rich RR, Fraser JD, Mariuzza RA, & Lee HK (**2001**) "Superantigen recognition by gammadelta T cells: SEA recognition site for human Vgamma2 T cell receptors." *Immunity (14:331-344)*:URL. See p. 137.

Morita CT, Jin C, Sarikonda G, & Wang H (**2007**) "Nonpeptide antigens, presentation mechanisms, and immunological memory of human Vgamma2Vdelta2 T cells: discriminating friend from foe through the recognition of prenyl pyrophosphate antigens." *Immunol Rev (215:59-76)*:DOI. See pp. 13, 27.

Morita M, Motoki K, Akimoto K, Natori T, Sakai T, Sawa E, Yamaji K, Koezuka Y, Kobayashi E, & Fukushima H (**1995 b**) "Structure-Activity Relationship of Alpha-Galactosylceramides against B16-Bearing Mice." *J Med Chem (38:2176-2187)*:URL. See p. 71.

Moses JE & Moorhouse AD (**2007**) "The growing applications of click chemistry." *Chem Soc Rev (36:1249-1262)*:URL. See p. 73.

Moulton KS, Vakili K, Zurakowski D, Soliman M, Butterfield C, Sylvin E, Lo KM, Gillies S, Javaherian K, & Folkman J (**2003**) "Inhibition of plaque neovascularization reduces macrophage accumulation and progression of advanced atherosclerosis." *Proc Natl Acad Sci U S A (100:4736-4741)*:URL. See p. 100.

Mukherjee S, Soe TT, & Maxfield FR (**1999**) "Endocytic sorting of lipid analogues differing solely in the chemistry of their hydrophobic tails." *J Cell Biol (144:1271-1284)*:URL. See pp. 16, 20.

Mukherjee S & Maxfield FR (**2000**) "Role of membrane organization and membrane domains in endocytic lipid trafficking." *Traffic (1:203-211)*:URL. See p. 134.

Munk ME, Gatrill AJ, & Kaufmann SH (**1990**) "Target cell lysis and IL-2 secretion by gamma/delta T lymphocytes after activation with bacteria." *J Immunol (145:2434-2439)*:URL. See p. 21.

Murata K, Toba T, Nakanishi K, Takahashi B, Yamamura T, Miyake S, & Annoura H (**2005**) "Total synthesis of an immunosuppressive glycolipid, (2S,3S,4R)-1-O- (alpha-d-galactosyl)-2-tetracosanoylamino-1,3,4-nonanetriol." *J Org Chem (70:2398-2401)*:URL. See p. 71.

Murphy S, Martin S, & Parton RG (**2009**) "Lipid droplet-organelle interactions; sharing the fats." *Biochim Biophys Acta (1791:441-447)*:DOI. See p. 10.

Mutschelknauss M, Kummer M, Muser J, Feinstein SB, Meyer PM, & Biedermann BC (**2007**) "Individual assessment of arteriosclerosis by empiric clinical profiling." *PLoS One (2:e1215)*:URL. See p. 92.

Naghavi M, Libby P, Falk E, Casscells SW, Litovsky S, Rumberger J, Badimon JJ, Stefanadis C, Moreno P, Pasterkamp G, Fayad Z, Stone PH, Waxman S, Raggi P, Madjid M, Zarrabi A, Burke A, Yuan C, Fitzgerald PJ, Siscovick DS, de Korte CL, Aikawa M, Airaksinen KE, Assmann G, Becker CR, Chesebro JH, Farb A, Galis ZS, Jackson C, Jang IK, Koenig W, Lodder RA, March K, Demirovic J, Navab M, Priori SG, Rekhter MD, Bahr R, Grundy SM, Mehran R, Colombo A, Boerwinkle E, Ballantyne C, Insull WJ, Schwartz RS, Vogel R, Serruys PW, Hansson GK, Faxon DP, Kaul S, Drexler H, Greenland P, Muller JE, Virmani R, Ridker PM, Zipes DP, Shah PK, & Willerson JT (**2003 a**) "From vulnerable plaque to vulnerable patient: a call for new definitions and risk assessment strategies: Part I." *Circulation (108:1664–1672)*:URL. See p. 89.

Naghavi M, Libby P, Falk E, Casscells SW, Litovsky S, Rumberger J, Badimon JJ, Stefanadis C, Moreno P, Pasterkamp G, Fayad Z, Stone PH, Waxman S, Raggi P, Madjid M, Zarrabi A, Burke A, Yuan C, Fitzgerald PJ, Siscovick DS, de Korte CL, Aikawa M, Airaksinen KE, Assmann G, Becker CR, Chesebro JH, Farb A, Galis ZS, Jackson C, Jang IK, Koenig W, Lodder RA, March K, Demirovic J, Navab M, Priori SG, Rekhter MD, Bahr R, Grundy SM, Mehran R, Colombo A, Boerwinkle E, Ballantyne C, Insull WJ, Schwartz RS, Vogel R, Serruys PW, Hansson GK, Faxon DP, Kaul S, Drexler H, Greenland P, Muller JE, Virmani R, Ridker PM, Zipes DP, Shah PK, & Willerson JT (**2003 b**) "From vulnerable plaque to vulnerable patient: a call for new definitions and risk assessment strategies: Part II." *Circulation (108:1772–1778)*:URL. See p. 89.

Naidenko OV, Maher JK, Ernst WA, Sakai T, Modlin RL, & Kronenberg M (**1999**) "Binding and antigen presentation of ceramide-containing glycolipids by soluble mouse and human CD1d molecules." *J Exp Med (190:1069–1080)*:URL. See p. 61.

Nakai Y, Iwabuchi K, Fujii S, Ishimori N, Dashtsoodol N, Watano K, Mishima T, Iwabuchi C, Tanaka S, Bezbradica JS, Nakayama T, Taniguchi M, Miyake S, Yamamura T, Kitabatake A, Joyce S, Van Kaer L, & Onoe K (**2004**) "Natural killer T cells accelerate atherogenesis in mice." *Blood (104:2051–2059)*:URL. See pp. 89, 102.

Naslavsky N, Weigert R, & Donaldson JG (**2003**) "Convergence of non-clathrin- and clathrin-derived endosomes involves Arf6 inactivation and changes in phosphoinositides." *Mol Biol Cell (14:417–431)*:DOI. See p. 119.

Ndonye RM, Izmirian DP, Dunn MF, Yu KOA, Porcelli SA, Khurana A, Kronenberg M, Richardson SK, & Howell AR (**2005**) "Synthesis and evaluation of sphinganine analogues of KRN7000 and OCH." *J Org Chem (70:10260–10270)*:URL. See p. 71.

Neefjes JJ, Momburg F, & Haemmerling GJ (**1993**) "Selective and ATP-dependent translocation of peptides by the MHC-encoded transporter." *Science (261:769–771)*:URL. See p. 148.

Neumann J, Schach N, & Koch N (**2001**) "Glycosylation signals that separate the trimerization from the mhc class II-binding domain control intracellular degradation of invariant chain." *J Biol Chem (276:13469–13475)*:DOI. See p. 108.

Newby AC (**2006**) "Do metalloproteinases destabilize vulnerable atherosclerotic plaques?" *Curr Opin Lipidol (17:556–561)*:URL. See p. 89.

Nguyen TKA, Koets AP, Santema WJ, van Eden W, Rutten VPMG, & Van Rhijn I (**2009**) "The mycobacterial glycolipid glucose monomycolate induces a memory T cell response comparable to a model protein antigen and no B cell response upon experimental vaccination of cattle." *Vaccine (27:4818–4825)*:DOI. See p. 7.

Niazi K, Chiu M, Mendoza R, Degano M, Khurana S, Moody D, Melian A, Wilson I, Kronenberg M, Porcelli S, & Modlin R (**2001**) "The A' and F' pockets of human CD1b are both required for optimal presentation of lipid antigens to T cells." *J Immunol (166:2562–2570)*:URL. See p. 18.

Niazi KR, Porcelli SA, & Modlin RL (**2002**) "The CD1b structure: antigen presentation adapts to a high-fat diet." *Nat Immunol (3:703–704)*:DOI. See p. 153.

Nowbakht P, Ionescu MCS, Rohner A, Kalberer CP, Rossy E, Mori L, Cosman D, De Libero G, & Wodnar-Filipowicz A (**2005**) "Ligands for natural killer cell-activating receptors are expressed upon the maturation of normal myelomonocytic cells but at low levels in acute myeloid leukemias." *Blood (105:3615–3622)*:DOI. See pp. 37, 38.

Nussbaumer P (**2008**) "Medicinal chemistry aspects sphingolipid metabolism." *ChemMedChem (3:543–551)*:URL. See p. 71.

Obeid LM, Linardic CM, Karolak LA, & Hannun YA (**1993**) "Programmed Cell-Death Induced by Ceramide." *Science (259:1769–1771)*:URL. See pp. 71, 73, 76.

Ono N, Van der Heijden I, Scheffer GL, Van de Wetering K, Van Deemter E, De Haas M, Boerke A, Gadella BM, De Rooij DG, Neefjes JJ, Groothuis TAM, Oomen L, Brocks L, Ishikawa T, & Borst P (**2007**) "Multidrug resistance-associated protein 9 (ABCC12) is present in mouse and boar sperm." *Biochem J (406:31–40)*:DOI. See p. 140.

Pachamuthu K & Schmidt RR (**2006**) "Synthetic routes to thiooligosaccharides and thioglycopeptides." *Chem Rev (106:160–187)*:DOI. See p. 71.

Paddison PJ, Caudy AA, Bernstein E, Hannon GJ, & Conklin DS (**2002**) "Short hairpin RNAs (shRNAs) induce sequence-specific silencing in mammalian cells." *Genes Dev (16:948–958)*:DOI. See p. 142.

Padron JM (**2006**) "Sphingolipids in anticancer therapy." *Curr Med Chem (13:755–770)*:URL. See pp. 73, 76.

Palmer E & Naeher D (**2009**) "Affinity threshold for thymic selection through a T-cell receptor-co-receptor zipper." *Nat Rev Immunol (9:207–213)*:DOI. See p. 69.

Palmer MH, Gaskell AJ, & Findlay RH (**1974**) "Molecular Energy-Levels of Azines - Ab-Initio Calculations and Correlation with Photoelectron-Spectroscopy." *J Chem Soc, Perkin Trans 2 (778–784)*:DOI. See p. 74.

Park SH & Bendelac A (**2000**) "CD1-restricted T-cell responses and microbial infection." *Nature (406:788–792)*:DOI. See p. 23.

Park YK, Lee JW, Ko YG, Hong S, & Park SH (**2005**) "Lipid rafts are required for efficient signal transduction by CD1d." *Biochem Biophys Res Commun (327:1143–1154)*:DOI. See pp. 107, 153.

Parker CM, Groh V, Band H, Porcelli SA, Morita C, Fabbi M, Glass D, Strominger JL, & Brenner MB (**1990**) "Evidence for extrathymic changes in the T cell receptor gamma/delta repertoire" *J Exp Med (171:1597–1612)*:URL. See p. 137.

Paul S, Breuninger LM, & Kruh GD (**1996**) "ATP-dependent transport of lipophilic cytotoxic drugs by membrane vesicles prepared from MRP-overexpressing HL60/ADR cells." *Biochemistry (35:14003–14011)*:DOI. See p. 139.

Paulsson G, Zhou X, Tornquist E, & Hansson GK (**2000**) "Oligoclonal T cell expansions in atherosclerotic lesions of apolipoprotein E-deficient mice." *Arterioscler Thromb Vasc Biol (20:10–17)*:URL. See p. 100.

Pawar DM, Khalil AA, Hooks DR, Collins K, Elliott T, Stafford J, Smith L, & Noe EA (**1998**) "E and Z conformations of esters, thiol esters, and amides." *J Am Chem Soc (120:2108–2112)*:DOI. See p. 80.

Pena-Cruz V, Ito S, Oukka M, Yoneda K, Dascher CC, Von Lichtenberg F, & Sugita M (**2001**) "Extraction of human Langerhans cells: a method for isolation of epidermis-resident dendritic cells." *J Immunol Methods (255:83–91)*:DOI. See p. 18.

Pena-Cruz V, Ito S, Dascher CC, Brenner MB, & Sugita M (**2003**) "Epidermal Langerhans cells efficiently mediate CD1a-dependent presentation of microbial lipid antigens to T cells." *J Invest Dermatol (121:517–521)*:DOI. See p. 18.

Pettus BJ, Chalfant CE, & Hannun YA (**2002**) "Ceramide in apoptosis: an overview and current perspectives." *Biochim Biophys Acta (1585:114–125)*:URL. See pp. 71, 73, 76.

Pfeffer K, Schoel B, Gulle H, Kaufmann SH, & Wagner H (**1990**) "Primary responses of human T cells to mycobacteria: a frequent set of gamma/delta T cells are stimulated by protease-resistant ligands." *Eur J Immunol (20:1175–1179)*:URL. See p. 138.

Pierre P, Turley SJ, Gatti E, Hull M, Meltzer J, Mirza A, Inaba K, Steinman RM, & Mellman I (**1997**) "Developmental regulation of MHC class II transport in mouse dendritic cells." *Nature (388:787–792)*:DOI. See p. 107.

Plettenburg O, Bodmer-Narkevitch V, & Wong CH (**2002**) "Synthesis of alpha-galactosyl ceramide, a potent immunostimulatory agent." *J Org Chem (67:4559–4564)*:URL. See pp. 73, 76.

Poloso NJ & Roche PA (**2004**) "Association of MHC class II-peptide complexes with plasma membrane lipid microdomains." *Curr Opin Immunol (16:103–107)*:URL. See p. 107.

Poloso NJ, Denzin LK, & Roche PA (**2006**) "CDw78 defines MHC class II-peptide complexes that require Ii chain-dependent lysosomal trafficking, not localization to a specific tetraspanin membrane microdomain." *J Immunol (177:5451–5458)*:URL. See pp. 110, 129.

Porcelli SA, Morita CT, & Modlin RL (**1996**) "T-cell recognition of non-peptide antigens." *Curr Opin Immunol (8:510–516)*:DOI. See p. 23.

Porcelli SA & Modlin RL (**1999**) "The CD1 system: antigen-presenting molecules for T cell recognition of lipids and glycolipids." *Annu Rev Immunol (17:297–329)*:DOI. See pp. 18, 23, 107.

Porcelli S, Brenner MB, Greenstein JL, Balk SP, Terhorst C, & Bleicher PA (**1989**) "Recognition of cluster of differentiation 1 antigens by human CD4-CD8-cytolytic T lymphocytes." *Nature (341:447–450)*:DOI. See pp. 3, 23.

Porcelli S, Morita CT, & Brenner MB (**1992**) "CD1b restricts the response of human CD4-8- T lymphocytes to a microbial antigen." *Nature (360:593–597)*:DOI. See pp. 3, 13.

Porcelli S, Yockey CE, Brenner MB, & Balk SP (**1993**) "Analysis of T cell antigen receptor (TCR) expression by human peripheral blood CD4-8- alpha/beta T cells demonstrates preferential use of several V beta genes and an invariant TCR alpha chain." *J Exp Med (178:1–16)*:URL. See p. 64.

Porubsky S, Speak AO, Luckow B, Cerundolo V, Platt FM, & Groene HJ (**2007**) "Normal development and function of invariant natural killer T cells in mice with isoglobotrihexosylceramide (iGb3) deficiency." *Proc Natl Acad Sci U S A (104:5977–5982)*:DOI. See p. 24.

Pratt S, Shepard RL, Kandasamy RA, Johnston PA, Perry W, & Dantzig AH (**2005**) "The multidrug resistance protein 5 (ABCC5) confers resistance to 5-fluorouracil and transports its monophosphorylated metabolites." *Mol Cancer Ther (4:855–863)*:DOI. See p. 153.

Prigozy TI, Sieling PA, Clemens D, Stewart PL, Behar SM, Porcelli SA, Brenner MB, Modlin RL, & Kronenberg M (**1997**) "The mannose receptor delivers lipoglycan antigens to endosomes for presentation to T cells by CD1b molecules." *Immunity (6:187–197)*:URL. See p. 134.

Purbhoo MA, Boulter JM, Price DA, Vuidepot AL, Hourigan CS, Dunbar PR, Olson K, Dawson SJ, Phillips RE, Jakobsen BK, Bell JI, & Sewell AK (**2001**) "The human CD8 coreceptor effects cytotoxic T cell activation and antigen sensitivity primarily by mediating complete phosphorylation of the T cell receptor zeta chain." *J Biol Chem (276:32786–32792)*:DOI. See p. 70.

Purcell WP & Singer JA (**1967**) "Electronic and Molecular Structure of Selected Unsubstituted and Dimethyl Amides from Measurements of Electric Moments and Nuclear Magnetic Resonance." *J Phys Chem (71:4316–4319)*:DOI. See p. 74.

Qian G, Qin X, Zang YQ, Ge B, Guo TB, Wan B, Fang L, & Zhang JZ (**2010**) "High doses of alpha-galactosylceramide potentiate experimental autoimmune encephalomyelitis by directly enhancing Th17 response." *Cell Res*:DOI. See p. 25.

Radhakrishna H & Donaldson JG (**1997**) "ADP-ribosylation factor 6 regulates a novel plasma membrane recycling pathway." *J Cell Biol (139:49–61)*:URL. See p. 119.

Radin NS (**2003**) "Designing anticancer drugs via the achilles heel: Ceramide, allylic ketones, and mitochondria." *Bioorg Med Chem (11:2123–2142)*:URL. See p. 71.

Rajan R, Wallimann K, Vasella A, Pace D, Genazzani AA, Canonico PL, & Condorelli F (**2004**) "Synthesis of 7-oxasphingosine and -ceramide analogues and their evaluation in a model for apoptosis." *Chemistry & Biodiversity (1:1785–1799)*:URL. See pp. 74, 76.

Rajan R, Mathew T, Buffa R, Bornancin F, Cavallari M, Nussbaumer P, De Libero G, & Vasella A (**2009**) "Synthesis and Evaluation of N-Acetyl-2-amino-2-deoxy-alpha-D-galactosyl 1-Thio-7-oxaceramide, a New Analogue of alpha-D-Galactosyl Ceramide." *Helv Chim Acta (92:918–927)*:DOI. See pp. 36, 70, 71.

Rasband WS (**1997–2009**) "ImageJ, U. S. National Institutes of Health, Bethesda, Maryland, USA, http://rsb.info.nih.gov/ij/":URL. See p. 55.

Ravna AW, Sylte I, & Sager G (**2008**) "A molecular model of a putative substrate releasing conformation of multidrug resistance protein 5 (MRP5)." *Eur J Med Chem (43:2557–2567)*:DOI. See p. 153.

Ravna AW & Sager G (**2009**) "Molecular modeling studies of ABC transporters involved in multidrug resistance." *Mini Rev Med Chem (9:186–193)*:URL. See p. 153.

Reape TJ & Groot PH (**1999**) "Chemokines and atherosclerosis." *Atherosclerosis (147:213–225)*:URL. See p. 103.

Reddy BG, Silk JD, Salio M, Balamurugan R, Shepherd D, Ritter G, Cerundolo V, & Schmidt RR. (**2009**) "Nonglycosidic agonists of invariant NKT cells for use as vaccine adjuvants." *ChemMedChem (4:171–175)*:DOI. See p. 153.

Redgrave JNE, Lovett JK, Gallagher PJ, & Rothwell PM (**2006**) "Histological assessment of 526 symptomatic carotid plaques in relation to the nature and timing of ischemic symptoms: the Oxford plaque study." *Circulation (113:2320–2328)*:URL. See p. 89.

Reid G, Wielinga P, Zelcer N, De Haas M, Van Deemter L, Wijnholds J, Balzarini J, & Borst P (**2003**) "Characterization of the transport of nucleoside analog drugs by the human multidrug resistance proteins MRP4 and MRP5." *Mol Pharmacol (63:1094–1103)*:URL. See p. 140.

Rensing L, Koch M, & Becker A (**2009**) "A comparative approach to the principal mechanisms of different memory systems." *Naturwissenschaften (96:1373–1384)*:DOI. See p. 7.

Resh MD (**1999**) "Fatty acylation of proteins: new insights into membrane targeting of myristoylated and palmitoylated proteins." *Biochim Biophys Acta (1451:1–16)*:URL. See p. 114.

Robertson AKL & Hansson GK (**2006**) "T cells in atherogenesis: for better or for worse?" *Arterioscler Thromb Vasc Biol (26:2421–2432)*:URL. See p. 89.

Rogers L, Burchat S, Gage J, Hasu M, Thabet M, Willcox L, Ramsamy TA, & Whitman SC (**2008**) "Deficiency of invariant V alpha 14 natural killer T cells decreases atherosclerosis in LDL receptor null mice." *Cardiovasc Res (78:167–174)*:URL. See p. 89.

Rohdich F, Bacher A, & Eisenreich W (**2005**) "Isoprenoid biosynthetic pathways as anti-infective drug targets." *Biochem Soc Trans (33:785–791)*:DOI. See p. 154.

Ross R (**1999**) "Atherosclerosis - an inflammatory disease." *N Engl J Med (340:115–126)*:URL. See p. 89.

Rostovtsev VV, Green LG, Fokin VV, & Sharpless KB (**2002**) "A stepwise Huisgen cycloaddition process: Copper(I)-catalyzed regioselective "ligation" of azides and terminal alkynes." *Angew Chem Int Ed Engl (41:2596–2599)*:URL. See p. 73.

Rovina P, Jaritz M, Hoefinger S, Graf C, Devay P, Billich A, Baumruker T, & Bornancin F (**2006**) "A critical beta6-beta7 loop in the pleckstrin homology domain of ceramide kinase." *Biochem J (400:255–265)*:DOI. See p. 73.

Rudd PM, Elliott T, Cresswell P, Wilson IA, & Dwek RA (**2001**) "Glycosylation and the immune system." *Science (291:2370–2376)*:URL. See p. 7.

Rudolph MG, Stanfield RL, & Wilson IA (**2006**) "How TCRs bind MHCs, peptides, and coreceptors." *Annu Rev Immunol (24:419–466)*:DOI. See pp. 24, 69.

Russano AM, Agea E, Corazzi L, Postle AD, De Libero G, Porcelli S, de Benedictis FM, & Spinozzi F (**2006**) "Recognition of pollen-derived phosphatidyl-ethanolamine by human CD1d-restricted gamma delta T cells." *J Allergy Clin Immunol (117:1178–1184)*:DOI. See p. 27.

Saada R, Weinberger M, Shahaf G, & Mehr R (**2007**) "Models for antigen receptor gene rearrangement: CDR3 length." *Immunol Cell Biol (85:323–332)*:DOI. See p. 69.

Sabharanjak S, Sharma P, Parton RG, & Mayor S (**2002**) "GPI-anchored proteins are delivered to recycling endosomes via a distinct cdc42-regulated, clathrin-independent pinocytic pathway." *Dev Cell (2:411–423)*:URL. See p. 114.

Sager G (**2004**) "Cyclic GMP transporters." *Neurochem Int (45:865–873)*:DOI. See p. 142.

Sager G & Ravna AW (**2009**) "Cellular efflux of cAMP and cGMP - a question about selectivity." *Mini Rev Med Chem (9:1009–1013)*:URL. See p. 153.

Sakai S, Mantani N, Kogure T, Ochiai H, Shimada Y, & Terasawa K (**2002**) "Gene expression of cell surface antigens in the early phase of murine influenza pneumonia determined by a cDNA expression array technique." *Mediators Inflamm (11:359–361)*:DOI. See p. 25.

Salamero J, Bausinger H, Mommaas AM, Lipsker D, Proamer F, Cazenave JP, Goud B, de la Salle H, & Hanau D (**2001**) "CD1a molecules traffic through the early recycling endosomal pathway in human Langerhans cells." *J Invest Dermatol (116:401–408)*:DOI. See pp. 113, 118.

Salomonsen J, Rathmann Sorensen M, Marston DA, Rogers SL, Collen T, van Hateren A, Smith AL, Beal RK, Skjodt K, & Kaufman J (**2005**) "Two CD1 genes map to the chicken MHC, indicating that CD1 genes are ancient and likely to have been present in the primordial MHC." *Proc Natl Acad Sci U S A (102:8668–8673)*:DOI. See p. 132.

Salter RD, Howell DN, & Cresswell P (**1985**) "Genes regulating HLA class I antigen expression in T-B lymphoblast hybrids." *Immunogenetics (21:235–246)*:URL. See p. 34.

Sandhoff K & Kolter T (**2003**) "Biosynthesis and degradation of mammalian glycosphingolipids." *Philos Trans R Soc Lond B Biol Sci (358:847–861)*:DOI. See p. 20.

Sandhoff R, Hepbildikler ST, Jennemann R, Geyer R, Gieselmann V, Proia RL, Wiegandt H, & Grone HJ (**2002**) "Kidney sulfatides in mouse models of inherited glycosphingolipid disorders: determination by nano-electrospray ionization tandem mass spectrometry." *J Biol Chem (277:20386–20398)*:DOI. See p. 125.

Sarikonda G, Wang H, Puan KJ, hui Liu X, Lee HK, Song Y, Distefano MD, Oldfield E, Prestwich GD, & Morita CT (**2008**) "Photoaffinity antigens for human gammadelta T cells." *J Immunol (181:7738–7750)*:URL. See p. 137.

Saudrais C, Spehner D, de la Salle H, Bohbot A, Cazenave JP, Goud B, Hanau D, & Salamero J (**1998**) "Intracellular pathway for the generation of functional MHC class II peptide complexes in immature human dendritic cells." *J Immunol (160:2597–2607)*:URL. See p. 131.

Savage AK, Constantinides MG, Han J, Picard D, Martin E, Li B, Lantz O, & Bendelac A (**2008**) "The transcription factor PLZF directs the effector program of the NKT cell lineage." *Immunity (29:391–403)*:DOI. See p. 24.

Savage PB, Teyton L, & Bendelac A (**2006**) "Glycolipids for natural killer T cells." *Chem Soc Rev (35:771–779)*:URL. See p. 74.

Schatz PJ (**1993**) "Use of peptide libraries to map the substrate specificity of a peptide-modifying enzyme: a 13 residue consensus peptide specifies biotinylation in Escherichia coli." *Biotechnology (N Y) (11:1138–1143)*:URL. See p. 37.

Scherer WF & Hoogasian AF (**1954**) "Preservation at subzero temperatures of mouse fibroblasts (strain L) and human epithelial cells (strain HeLa)." *Proc Soc Exp Biol Med (87:480–487)*:URL. See p. 34.

Schiefner A, Fujio M, Wu D, Wong CH, & Wilson IA (**2009**) "Structural evaluation of potent NKT cell agonists: implications for design of novel stimulatory ligands." *J Mol Biol (394:71–82)*:DOI. See p. 153.

Schlesinger LS, Kaufman TM, Iyer S, Hull SR, & Marchiando LK (**1996**) "Differences in mannose receptor-mediated uptake of lipoarabinomannan from virulent and attenuated strains of Mycobacterium tuberculosis by human macrophages." *J Immunol (157:4568–4575)*:URL. See p. 127.

Schoel B, Sprenger S, & Kaufmann SH (**1994**) "Phosphate is essential for stimulation of V gamma 9V delta 2 T lymphocytes by mycobacterial low molecular weight ligand." *Eur J Immunol (24:1886–1892)*:URL. See p. 138.

Schuemann J, Facciotti F, Panza L, Michieletti M, Compostella F, Collmann A, Mori L, & De Libero G (**2007**) "Differential alteration of lipid antigen presentation to NKT cells due to imbalances in lipid metabolism." *Eur J Immunol (37:1431–1441)*:URL. See pp. 20, 36.

Schulze H, Kolter T, & Sandhoff K (**2009**) "Principles of lysosomal membrane degradation: Cellular topology and biochemistry of lysosomal lipid degradation." *Biochim Biophys Acta (1793:674–683)*:DOI. See p. 20.

Schweizer A, Kornfeld S, & Rohrer J (**1996**) "Cysteine34 of the cytoplasmic tail of the cation-dependent mannose 6-phosphate receptor is reversibly palmitoylated and required for normal trafficking and lysosomal enzyme sorting." *J Cell Biol (132:577–584)*:URL. See p. 115.

Scotet E, Martinez LO, Grant E, Barbaras R, Jenoe P, Guiraud M, Monsarrat B, Saulquin X, Maillet S, Esteve JP, Lopez F, Perret B, Collet X, Bonneville M, & Champagne E (**2005**) "Tumor recognition following Vgamma9Vdelta2 T cell receptor interactions with a surface F1-ATPase-related structure and apolipoprotein A-I." *Immunity (22:71–80)*:DOI. See p. 21.

Scott-Ward TS, Li H, Schmidt A, Cai Z, & Sheppard DN (**2004**) "Direct block of the cystic fibrosis transmembrane conductance regulator Cl(-) channel by niflumic acid." *Mol Membr Biol (21:27–38)*:DOI. See p. 140.

Sengupta P, Baird B, & Holowka D (**2007**) "Lipid rafts, fluid/fluid phase separation, and their relevance to plasma membrane structure and function." *Semin Cell Dev Biol (18:583–590)*:DOI. See pp. 12, 13.

Setterblad N, Roucard C, Bocaccio C, Abastado JP, Charron D, & Mooney N (**2003**) "Composition of MHC class II-enriched lipid microdomains is modified during maturation of primary dendritic cells." *J Leukoc Biol (74:40–48)*:URL. See p. 129.

Shamshiev A, Donda A, Carena I, Mori L, Kappos L, & De Libero G (**1999**) "Self glycolipids as T-cell autoantigens." *Eur J Immunol (29:1667–1675)*:URL. See p. 35.

Shamshiev A, Donda A, Prigozy TI, Mori L, Chigorno V, Benedict CA, Kappos L, Sonnino S, Kronenberg M, & De Libero G (**2000**) "The alphabeta T cell response to self-glycolipids shows a novel mechanism of CD1b loading and a requirement for complex oligosaccharides." *Immunity (13:255–264)*:URL. See pp. 20, 42, 134.

Shamshiev A, Gober HJ, Donda A, Mazorra Z, Mori L, & De Libero G (**2002**) "Presentation of the same glycolipid by different CD1 molecules." *J Exp Med (195:1013–1021)*:URL. See pp. 35, 81, 117, 125, 132.

Shaw S, Ziegler A, & DeMars R (**1985**) "Specificity of monoclonal antibodies directed against human and murine class II histocompatibility antigens as analyzed by binding to HLA-deletion mutant cell lines." *Hum Immunol (12:191–211)*:URL. See p. 33.

Shikata K, Niiro H, Azuma H, Ogino K, & Tachibana T (**2003**) "Apoptotic activities of C2-ceramide and C2-dihydroceramide homologues against HL-60 cells." *Bioorg Med Chem (11:2723–2728)*:URL. See pp. 71, 73, 76.

Shimamura M, Huang YY, Okamoto N, Suzuki N, Yasuoka J, Morita K, Nishiyama A, Amano Y, & Mishina T (**2007**) "Modulation of V alpha 19 NKT cell immune responses by alpha-mannosyl ceramide derivatives consisting of a series of modified sphingosines." *Eur J Immunol (37:1836–1844)*:URL. See p. 71.

Shnyrova A, Frolov VA, & Zimmerberg J (**2008**) "ER biogenesis: self-assembly of tubular topology by protein hairpins." *Curr Biol (18:R474–R476)*:DOI. See p. 148.

Sieling PA, Chatterjee D, Porcelli SA, Prigozy TI, Mazzaccaro RJ, Soriano T, Bloom BR, Brenner MB, Kronenberg M, & Brennan PJ (**1995**) "CD1-restricted T cell recognition of microbial lipoglycan antigens." *Science (269:227–230)*:URL. See p. 14.

Sieling PA, Ochoa MT, Jullien D, Leslie DS, Sabet S, Rosat JP, Burdick AE, Rea TH, Brenner MB, Porcelli SA, & Modlin RL (**2000**) "Evidence for human CD4+ T cells in the CD1-restricted repertoire: derivation of mycobacteria-reactive T cells from leprosy lesions." *J Immunol (164:4790–4796)*:URL. See p. 23.

Silk JD, Hermans IF, Gileadi U, Chong TW, Shepherd D, Salio M, Mathew B, Schmidt RR, Lunt SJ, Williams KJ, Stratford IJ, Harris AL, & Cerundolo V (**2004**) "Utilizing the adjuvant properties of CD1d-dependent NK T cells in T cell-mediated immunotherapy." *J Clin Invest (114:1800–1811)*:DOI. See p. 153.

Silk JD, Salio M, Reddy BG, Shepherd D, Gileadi U, Brown J, Masri SH, Polzella P, Ritter G, Besra GS, Jones EY, Schmidt RR, & Cerundolo V (**2008 a**) "Cutting edge: nonglycosidic CD1d lipid ligands activate human and murine invariant NKT cells." *J Immunol (180:6452–6456)*:URL. See p. 69.

Silk JD, Salio M, Brown J, Jones EY, & Cerundolo V (**2008 b**) "Structural and functional aspects of lipid binding by CD1 molecules." *Annu Rev Cell Dev Biol (24:369–395)*:DOI. See pp. 20, 69.

Simonini A, Moscucci M, Muller DW, Bates ER, Pagani FD, Burdick MD, & Strieter RM (**2000**) "IL-8 is an angiogenic factor in human coronary atherectomy tissue." *Circulation (101:1519–1526)*:URL. See pp. 97, 103.

Simons K & Ikonen E (**1997**) "Functional rafts in cell membranes." *Nature (387:569–572)*:DOI. See pp. 12, 110.

Sloma I, Zilber MT, Charron D, Girot R, Tamouza R, & Gelin C (**2004**) "Upregulation and atypical expression of the CD1 molecules on monocytes in sickle cell disease." *Hum Immunol (65:1370–1376)*:DOI. See p. 107.

Sloma I, Zilber MT, Vasselon T, Setterblad N, Cavallari M, Mori L, De Libero G, Charron D, Mooney N, & Gelin C (**2008**) "Regulation of CD1a surface expression and antigen presentation by invariant chain and lipid rafts." *J Immunol (180:980–987)*:URL. See pp. 107, 108, 110, 111, 128, 132.

Snapp EL, Hegde RS, Francolini M, Lombardo F, Colombo S, Pedrazzini E, Borgese N, & Lippincott-Schwartz J (**2003**) "Formation of stacked ER cisternae by low affinity protein interactions." *J Cell Biol (163:257–269)*:DOI. See p. 148.

Snook CF, Jones JA, & Hannun YA (**2006**) "Sphingolipid-binding proteins." *Biochim Biophys Acta Mol Cell Biol Lipids (1761:927–946)*:URL. See p. 73.

Somerharju P, Virtanen JA, Cheng KH, & Hermansson M (**2009**) "The superlattice model of lateral organization of membranes and its implications on membrane lipid homeostasis." *Biochim Biophys Acta (1788:12–23)*:DOI. See p. 9.

Spada FM, Grant EP, Peters PJ, Sugita M, Melian A, Leslie DS, Lee HK, van Donselaar E, Hanson DA, Krensky AM, Majdic O, Porcelli SA, Morita CT, & Brenner MB (**2000**) "Self-recognition of CD1 by gamma/delta T cells: implications for innate immunity." *J Exp Med (191:937–948)*:URL. See p. 27.

Speak AO, Salio M, Neville DCA, Fontaine J, Priestman DA, Platt N, Heare T, Butters TD, Dwek RA, Trottein F, Exley MA, Cerundolo V, & Platt FM (**2007**) "Implications for invariant natural killer T cell ligands due to the restricted presence of isoglobotrihexosylceramide in mammals." *Proc Natl Acad Sci U S A (104:5971–5976)*:DOI. See p. 24.

Spiegel S & Milstien S (**2007**) "Functions of the multifaceted family of sphingosine kinases and some close relatives." *J Biol Chem (282:2125–2129)*:URL. See pp. 72, 81.

Staros JV, Bayley H, Standring DN, & Knowles JR (**1978**) "Reduction of aryl azides by thiols: implications for the use of photoaffinity reagents." *Biochem Biophys Res Commun (80:568–572)*:URL. See p. 72.

Stary HC, Blankenhorn DH, Chandler AB, Glagov S, Insull W, Richardson M, Rosenfeld ME, Schaffer SA, Schwartz CJ, & Wagner WD (**1992**) "A definition of the intima of human arteries and of its atherosclerosis-prone regions. A report from the Committee on Vascular Lesions of the Council on Arteriosclerosis, American Heart Association." *Circulation (85:391–405)*:URL. See p. 90.

Steg PG, Bhatt DL, Wilson PWF, D'Agostino RS, Ohman EM, Roether J, Liau CS, Hirsch AT, Mas JL, Ikeda Y, Pencina MJ, & Goto S (**2007**) "One-year cardiovascular event rates in outpatients with atherothrombosis." *JAMA (297:1197–1206)*:URL. See p. 89.

Stenger S, Niazi KR, & Modlin RL (**1998**) "Down-regulation of CD1 on antigen-presenting cells by infection with Mycobacterium tuberculosis." *J Immunol (161:3582–3588)*:URL. See p. 153.

Storkus WJ, Alexander J, Payne JA, Dawson JR, & Cresswell P (**1989**) "Reversal of natural killing susceptibility in target cells expressing transfected class I HLA genes." *Proc Natl Acad Sci U S A (86:2361–2364)*:URL. See p. 34.

Sugita M, Jackman RM, van Donselaar E, Behar SM, Rogers RA, Peters PJ, Brenner MB, & Porcelli SA (**1996**) "Cytoplasmic tail-dependent localization of CD1b antigen-presenting molecules to MIICs." *Science (273:349–352)*:URL. See pp. 18, 19, 113.

Sugita M, Grant EP, van Donselaar E, Hsu VW, Rogers RA, Peters PJ, & Brenner MB (**1999**) "Separate pathways for antigen presentation by CD1 molecules." *Immunity (11:743–752)*:URL. See pp. 19, 113, 115, 118, 119, 125, 131.

Sugita M, van Der Wel N, Rogers RA, Peters PJ, & Brenner MB (**2000 *a***) "CD1c molecules broadly survey the endocytic system." *Proc Natl Acad Sci U S A (97:8445–8450)*:DOI. See p. 19.

Sugita M, Peters PJ, & Brenner MB (**2000 *b***) "Pathways for lipid antigen presentation by CD1 molecules: nowhere for intracellular pathogens to hide." *Traffic (1:295–300)*:URL. See p. 20.

Sugita M, Barral DC, & Brenner MB (**2007**) "Pathways of CD1 and lipid antigen delivery, trafficking, processing, loading, and presentation." *Curr Top Microbiol Immunol (314:143–164)*:URL. See p. 20.

Sugita M, Cao X, Watts GFM, Rogers RA, Bonifacino JS, & Brenner MB (**2002**) "Failure of trafficking and antigen presentation by CD1 in AP-3-deficient cells." *Immunity (16:697–706)*:URL. See pp. 19, 133.

Suzuki T, Sasaki H, Kuh HJ, Agui M, Tatsumi Y, Tanabe S, Terada M, Saijo N, & Nishio K (**2000**) "Detailed structural analysis on both human MRP5 and mouse mrp5 transcripts." *Gene (242:167–173)*:URL. See p. 153.

Tabas I, Williams KJ, & Boren J (**2007**) "Subendothelial lipoprotein retention as the initiating process in atherosclerosis: update and therapeutic implications." *Circulation (116:1832–1844)*:URL. See p. 89.

Taghon T & Rothenberg EV (**2008**) "Molecular mechanisms that control mouse and human TCR-alphabeta and TCR-gammadelta T cell development." *Semin Immunopathol (30:383–398)*:DOI. See p. 26.

Tan RX & Chen JH (**2003**) "The cerebrosides." *Natural Product Reports (20:509–534)*:URL. See p. 71.

Tanaka Y, Sano S, Nieves E, De Libero G, Rosa D, Modlin RL, Brenner MB, Bloom BR, & Morita CT (**1994**) "Nonpeptide ligands for human gamma delta T cells." *Proc Natl Acad Sci U S A (91:8175–8179)*:URL. See pp. 21, 138.

Tanaka Y, Morita CT, Nieves E, Brenner MB, & Bloom BR (**1995**) "Natural and synthetic non-peptide antigens recognized by human gamma delta T cells." *Nature (375:155–158)*:DOI. See pp. 13, 137.

Taniguchi M, Tashiro T, Dashtsoodol N, Hongo N, & Watarai H (**2010**) "The specialized iNKT cell system recognizes glycolipid antigens and bridges the innate and acquired immune systems with potential applications for cancer therapy." *Int Immunol (22:1–6)*:DOI. See pp. 25, 84.

Tarrant JM, Robb L, van Spriel AB, & Wright MD (**2003**) "Tetraspanins: molecular organisers of the leukocyte surface." *Trends Immunol (24:610–617)*:URL. See p. 129.

Tashiro T, Nakagawa R, Inoue S, Shiozaki M, Watarai H, Taniguchi M, & Mori K (**2008**) "RCAI-61, the 6'-O-methylated analog of KRN7000: its synthesis and potent bioactivity for mouse lymphocytes to produce interferon-gamma in vivo." *Tetrahedron Letters (49:6827–6830)*:DOI. See p. 71.

Terabe M & Berzofsky JA (**2008**) "The role of NKT cells in tumor immunity." *Adv Cancer Res (101:277–348)*:DOI. See p. 23.

Terkeltaub R, Banka CL, Solan J, Santoro D, Brand K, & Curtiss LK (**1994**) "Oxidized LDL induces monocytic cell expression of interleukin-8, a chemokine with T-lymphocyte chemotactic activity." *Arterioscler Thromb (14:47–53)*:URL. See p. 103.

Thedrez A, Sabourin C, Gertner J, Devilder MC, Allain-Maillet S, Fournie JJ, Scotet E, & Bonneville M (**2007**) "Self/non-self discrimination by human gammadelta T cells: simple solutions for a complex issue?" *Immunol Rev (215:123–135)*:DOI. See p. 27.

Thevissen K, Francois I, Winderickx J, Pannecouque C, & Cammue BPA (**2006**) "Ceramide involvement in apoptosis and apoptotic diseases." *Mini Rev Med Chem (6:699–709)*:URL. See p. 81.

Thompson K, Rojas-Navea J, & Rogers MJ (**2006**) "Alkylamines cause Vgamma9Vdelta2 T-cell activation and proliferation by inhibiting the mevalonate pathway." *Blood (107:651–654)*:DOI. See p. 13.

Timmer RT & Gunn RB (**1998**) "Phosphate transport by the human renal cotransporter NaPi-3 expressed in HEK-293 cells." *Am J Physiol (274:C757–C769)*:URL. See p. 138.

To K, Agrotis A, Besra G, Bobik A, & Toh BH (**2009**) "NKT cell subsets mediate differential proatherogenic effects in ApoE-/- mice." *Arterioscler Thromb Vasc Biol (29:671–677)*:DOI. See p. 102.

Torky ARW, Stehfest E, Viehweger K, Taege C, & Foth H (**2005**) "Immuno-histochemical detection of MRPs in human lung cells in culture." *Toxicology (207:437–450)*:DOI. See p. 148.

Tough DF & Sprent J (**1998**) "Lifespan of gamma/delta T cells" *J Exp Med (187:357–365)*:URL. See p. 26.

Tsuchiya S, Yamabe M, Yamaguchi Y, Kobayashi Y, Konno T, & Tada K (**1980**) "Establishment and characterization of a human acute monocytic leukemia cell line (THP-1)." *Int J Cancer (26:171–176)*:URL. See p. 34.

Tsuneyama K, Yasoshima M, Harada K, Hiramatsu K, Gershwin ME, & Nakanuma Y (**1998**) "Increased CD1d expression on small bile duct epithelium and epithelioid granulomas in livers in primary biliary cirrhosis." *Hepatology (28:620–623)*:DOI. See p. 18.

Tupin E, Kinjo Y, & Kronenberg M (**2007**) "The unique role of natural killer T cells in the response to microorganisms." *Nat Rev Microbiol (5:405–417)*:URL. See pp. 26, 71.

Tupin E, Nicoletti A, Elhage R, Rudling M, Ljunggren HG, Hansson GK, & Berne GP (**2004**) "CD1d-dependent activation of NKT cells aggravates atherosclerosis." *J Exp Med (199:417–422)*:URL. See pp. 89, 102.

Turley SJ, Inaba K, Garrett WS, Ebersold M, Unternaehrer J, Steinman RM, & Mellman I (**2000**) "Transport of peptide-MHC class II complexes in developing dendritic cells." *Science (288:522–527)*:URL. See p. 107.

Tusnady GE, Sarkadi B, Simon I, & Varadi A (**2006**) "Membrane topology of human ABC proteins." *FEBS Lett (580:1017–1022)*:DOI. See p. 140.

Tysoe-Calnon VA, Grundy JE, & Perkins SJ (**1991**) "Molecular comparisons of the beta 2-microglobulin-binding site in class I major-histocompatibility-complex alpha-chains and proteins of related sequences." *Biochem J (277 (Pt 2):359–369)*:URL. See p. 16.

Unternaehrer JJ, Chow A, Pypaert M, Inaba K, & Mellman I (**2007**) "The tetraspanin CD9 mediates lateral association of MHC class II molecules on the dendritic cell surface." *Proc Natl Acad Sci U S A (104:234–239)*:DOI. See pp. 108, 129.

Van Voorhis WC, Steinman RM, Hair LS, Luban J, Witmer MD, Koide S, & Cohn ZA (**1983**) "Specific antimononuclear phagocyte monoclonal antibodies. Application to the purification of dendritic cells and the tissue localization of macrophages." *J Exp Med (158:126–145)*:URL. See p. 33.

Van Wazer JR (**1964**) "The Chemistry Of Phosphates." *J Dent Res (43:SUPPL:1052–SUPPL:1064)*:URL. See p. 137.

van der Wal AC, Das PK, Bentz van de Berg D, van der Loos CM, & Becker AE (**1989**) "Atherosclerotic lesions in humans. In situ immunophenotypic analysis suggesting an immune mediated response." *Lab Invest (61:166–170)*:URL. See p. 89.

van der Wal AC, Becker AE, van der Loos CM, & Das PK (**1994**) "Site of intimal rupture or erosion of thrombosed coronary atherosclerotic plaques is characterized by an inflammatory process irrespective of the dominant plaque morphology." *Circulation (89:36–44)*:URL. See p. 89.

van der Wel NN, Sugita M, Fluitsma DM, Cao X, Schreibelt G, Brenner MB, & Peters PJ (**2003**) "CD1 and major histocompatibility complex II molecules follow a different course during dendritic cell maturation." *Mol Biol Cell (14:3378–3388)*:DOI. See pp. 18, 127.

van Meer G (**2005**) "Cellular lipidomics." *EMBO J (24:3159–3165)*:DOI. See p. 9.

van Meer G, Voelker DR, & Feigenson GW (**2008**) "Membrane lipids: where they are and how they behave." *Nat Rev Mol Cell Biol (9:112–124)*:DOI. See pp. 9, 11, 137, 153.

van Meer G & Hoetzl S (**2009**) "Sphingolipid topology and the dynamic organization and function of membrane proteins." *FEBS Lett*:DOI. See p. 10.

VanderLaan PA, Reardon CA, Sagiv Y, Blachowicz L, Lukens J, Nissenbaum M, Wang CR, & Getz GS (**2007**) "Characterization of the natural killer T-cell response in an adoptive transfer model of atherosclerosis." *Am J Pathol (170:1100–1107)*:URL. See p. 89.

Vankar YD & Schmidt RR (**2000**) "Chemistry of glycosphingolipids-carbohydrate molecules of biological significance." *Chem Soc Rev (29:201–216)*:DOI. See p. 71.

Vincent MS, Gumperz JE, & Brenner MB (**2003**) "Understanding the function of CD1-restricted T cells." *Nat Immunol (4:517–523)*:DOI. See p. 15.

Virmani R, Kolodgie FD, Burke AP, Finn AV, Gold HK, Tulenko TN, Wrenn SP, & Narula J (**2005**) "Atherosclerotic plaque progression and vulnerability to rupture: angiogenesis as a source of intraplaque hemorrhage." *Arterioscler Thromb Vasc Biol (25:2054–2061)*:URL. See p. 89.

Vos JP, Lopes-Cardozo M, & Gadella BM (**1994**) "Metabolic and functional aspects of sulfogalactolipids." *Biochim Biophys Acta (1211:125–149)*:URL. See p. 125.

Walgren RA, Karnaky KJ, Lindenmayer GE, & Walle T (**2000**) "Efflux of dietary flavonoid quercetin 4'-beta-glucoside across human intestinal Caco-2 cell monolayers by apical multidrug resistance-associated protein-2." *J Pharmacol Exp Ther (294:830–836)*:URL. See p. 140.

Wang Jh & Reinherz EL (**2002**) "Structural basis of T cell recognition of peptides bound to MHC molecules." *Mol Immunol (38:1039–1049)*:URL. See pp. 24, 69.

Wang N, Tabas I, Winchester R, Ravalli S, Rabbani LE, & Tall A (**1996**) "Interleukin 8 is induced by cholesterol loading of macrophages and expressed by macrophage foam cells in human atheroma." *J Biol Chem (271:8837–8842)*:URL. See p. 103.

Warmerdam PA, Long EO, & Roche PA (**1996**) "Isoforms of the invariant chain regulate transport of MHC class II molecules to antigen processing compartments." *J Cell Biol (133:281–291)*:URL. See p. 131.

Warren MS, Zerangue N, Woodford K, Roberts LM, Tate EH, Feng B, Li C, Feuerstein TJ, Gibbs J, Smith B, de Morais SM, Dower WJ, & Koller KJ (**2009**) "Comparative gene expression profiles of ABC transporters in brain microvessel endothelial cells and brain in five species including human." *Pharmacol Res (59:404–413)*:DOI. See p. 153.

Watarai H, Rybouchkin A, Hongo N, Nagata Y, Sakata S, Sekine E, Dashtsoodol N, Tashiro T, Fujii SI, Shimizu K, Mori K, Masuda K, Kawamoto H, Koseki H, & Taniguchi M (**2009**) "Generation of functional NKT cells in vitro from embryonic stem cells bearing rearranged invariant Valpha14-Jalpha18 TCRalpha gene." *Blood*:DOI. See p. 153.

Wei H, Huang D, Lai X, Chen M, Zhong W, Wang R, & Chen ZW (**2008**) "Definition of APC presentation of phosphoantigen (E)-4-hydroxy-3-methyl-but-2-enyl pyrophosphate to Vgamma2Vdelta 2 TCR." *J Immunol (181:4798–4806)*:URL. See p. 137.

Weigert MG, Cesari IM, Yonkovich SJ, & Cohn M (**1970**) "Variability in the lambda light chain sequences of mouse antibody." *Nature (228:1045–1047)*:URL. See p. 35.

Weigert R, Yeung AC, Li J, & Donaldson JG (**2004**) "Rab22a regulates the recycling of membrane proteins internalized independently of clathrin." *Mol Biol Cell (15:3758–3770)*:DOI. See pp. 118, 119.

Weinreich MA & Hogquist KA (**2008**) "Thymic emigration: when and how T cells leave home." *J Immunol (181:2265–2270)*:URL. See p. 7.

Whelan JA, Dunbar PR, Price DA, Purbhoo MA, Lechner F, Ogg GS, Griffiths G, Phillips RE, Cerundolo V, & Sewell AK (**1999**) "Specificity of CTL interactions with peptide-MHC class I tetrameric complexes is temperature dependent." *J Immunol (163:4342–4348)*:URL. See pp. 63, 69.

Wijnholds J, Evers R, van Leusden MR, Mol CA, Zaman GJ, Mayer U, Beijnen JH, van der Valk M, Krimpenfort P, & Borst P (**1997**) "Increased sensitivity to anticancer drugs and decreased inflammatory response in mice lacking the multidrug resistance-associated protein." *Nat Med (3:1275–1279)*:URL. See p. 145.

Willcox BE, Willcox CR, Dover LG, & Besra G (**2007**) "Structures and functions of microbial lipid antigens presented by CD1." *Curr Top Microbiol Immunol (314:73–110)*:URL. See p. 14.

Wooldridge L, Lissina A, Cole DK, van den Berg HA, Price DA, & Sewell AK (**2009**) "Tricks with tetramers: how to get the most from multimeric peptide-MHC." *Immunology (126:147–164)*:DOI. See pp. 61, 69, 70.

Wu CP, Woodcock H, Hladky SB, & Barrand MA (**2005**) "cGMP (guanosine 3',5'-cyclic monophosphate) transport across human erythrocyte membranes." *Biochem Pharmacol (69:1257–1262)*:URL. See p. 49.

Wu D, Fujio M, & Wong CH (**2008**) "Glycolipids as immunostimulating agents." *Bioorg Med Chem (16:1073–1083)*:URL. See pp. 73, 76.

Wu D, Zajonc DM, Fujio M, Sullivan BA, Kinjo Y, Kronenberg M, Wilson IA, & Wong CH (**2006**) "Design of natural killer T cell activators: structure and function of a microbial glycosphingolipid bound to mouse CD1d." *Proc Natl Acad Sci U S A (103:3972–3977)*:DOI. See pp. 18, 20.

Wun KS, Borg NA, Kjer-Nielsen L, Beddoe T, Koh R, Richardson SK, Thakur M, Howell AR, Scott-Browne JP, Gapin L, Godfrey DI, McCluskey J, & Rossjohn J (**2008**) "A minimal binding footprint on CD1d-glycolipid is a basis for selection of the unique human NKT TCR." *J Exp Med (205:939–949)*:DOI. See pp. 24, 69, 72, 82.

Xavier R & Seed B (**1999**) "Membrane compartmentation and the response to antigen." *Curr Opin Immunol (11:265–269)*:URL. See pp. 110, 129.

Xia C, Yao Q, Schuemann J, Rossy E, Chen W, Zhu L, Zhang W, De Libero G, & Wang PG (**2006**) "Synthesis and biological evaluation of alpha-galactosylceramide (KRN7000) and isoglobotrihexosylceramide (iGb3)." *Bioorg Med Chem Lett (16:2195–2199)*:URL. See p. 35.

Xiong N & Raulet DH (**2007**) "Development and selection of gammadelta T cells." *Immunol Rev (215:15–31)*:DOI. See p. 26.

Yamamoto T, Hasegawa H, Hakogi T, & Katsumura S (**2006**) "Versatile synthetic method for sphingolipids and functionalized sphingosine derivatives via olefin cross metathesis." *Org Lett (8:5569–5572)*:DOI. See p. 71.

Yang GL, Schmieg J, Tsuji M, & Franck RW (**2004**) "The C-glycoside analogue of the immunostimulant alpha-galactosylceramide (KRN7000): Synthesis and striking enhancement of activity." *Angew Chem Int Ed Engl (43:3818–3822)*:URL. See pp. 71, 73, 76.

Yuan W, Dasgupta A, & Cresswell P (**2006**) "Herpes simplex virus evades natural killer T cell recognition by suppressing CD1d recycling." *Nat Immunol (7:835–842)*:DOI. See p. 20.

Yuan W, Qi X, Tsang P, Kang SJ, Illaniorov PA, Besra GS, Gumperz J, & Cresswell P (**2007**) "Saposin B is the dominant saposin that facilitates lipid binding to human CD1d molecules." *Proc Natl Acad Sci U S A (104:5551–5556)*:DOI. See p. 20.

Yuan W, Kang SJ, Evans JE, & Cresswell P (**2009**) "Natural lipid ligands associated with human CD1d targeted to different subcellular compartments." *J Immunol (182:4784–4791)*:DOI. See pp. 20, 25, 82.

Yue TL, Wang X, Sung CP, Olson B, McKenna PJ, Gu JL, & Feuerstein GZ (**1994**) "Interleukin-8. A mitogen and chemoattractant for vascular smooth muscle cells." *Circ Res (75:1–7)*:URL. See p. 103.

Yusufi AN, Szczepanska-Konkel M, Kempson SA, McAteer JA, & Dousa TP (**1986**) "Inhibition of human renal epithelial Na+/Pi cotransport by phosphonoformic acid." *Biochem Biophys Res Commun (139:679–686)*:URL. See p. 138.

Yuyama K, Sekino-Suzuki N, Sanai Y, & Kasahara K (**2003**) "Lipid rafts in cellular signaling and disease." *Trends in Glycoscience and Glycotechnology (15:139–151)*:URL. See p. 12.

Zajonc DM & Wilson IA (**2007**) "Architecture of CD1 proteins." *Curr Top Microbiol Immunol (314:27–50)*:URL. See pp. 16, 68, 82.

Zajonc DM, Elsliger MA, Teyton L, & Wilson IA (**2003**) "Crystal structure of CD1a in complex with a sulfatide self antigen at a resolution of 2.15 A." *Nat Immunol (4:808–815)*:DOI. See p. 132.

Zajonc DM, Maricic I, Wu D, Halder R, Roy K, Wong CH, Kumar V, & Wilson IA (**2005 a**) "Structural basis for CD1d presentation of a sulfatide derived from myelin and its implications for autoimmunity." *J Exp Med (202:1517–1526)*:DOI. See pp. 25, 125.

Zajonc DM, Cantu C, Mattner J, Zhou D, Savage PB, Bendelac A, Wilson IA, & Teyton L (**2005 b**) "Structure and function of a potent agonist for the semi-invariant natural killer T cell receptor." *Nat Immunol (6:810–818)*:DOI. See pp. 18, 20.

Zajonc DM, Ainge GD, Painter GF, Severn WB, & Wilson IA (**2006**) "Structural characterization of mycobacterial phosphatidylinositol mannoside binding to mouse CD1d." *J Immunol (177:4577–4583)*:URL. See p. 14.

Zehmer JK, Huang Y, Peng G, Pu J, Anderson RGW, & Liu P (**2009**) "A role for lipid droplets in inter-membrane lipid traffic." *Proteomics (9:914–921)*:DOI. See p. 10.

Zeng Z, Castano AR, Segelke BW, Stura EA, Peterson PA, & Wilson IA (**1997**) "Crystal structure of mouse CD1: An MHC-like fold with a large hydrophobic binding groove." *Science (277:339–345)*:URL. See pp. 4, 16.

Zhou DP, Mattner J, Cantu C, Schrantz N, Yin N, Gao Y, Sagiv Y, Hudspeth K, Wu YP, Yamashita T, Teneberg S, Wang DC, Proia RL, Levery SB, Savage PB, Teyton L,

& Bendelac A (**2004 a**) "Lysosomal glycosphingolipid recognition by NKT cells." *Science (306:1786–1789)*:URL. See p. 24.

Zhou D, Cantu C r, Sagiv Y, Schrantz N, Kulkarni AB, Qi X, Mahuran DJ, Morales CR, Grabowski GA, Benlagha K, Savage P, Bendelac A, & Teyton L (**2004 b**) "Editing of CD1d-bound lipid antigens by endosomal lipid transfer proteins." *Science (303:523–527)*:URL. See p. 20.

Zhou SF, Wang LL, Di YM, Xue CC, Duan W, Li CG, & Li Y (**2008**) "Substrates and inhibitors of human multidrug resistance associated proteins and the implications in drug development." *Curr Med Chem (15:1981–2039)*:URL. See p. 140.

Zilber MT, Setterblad N, Vasselon T, Doliger C, Charron D, Mooney N, & Gelin C (**2005**) "MHC class II/CD38/CD9: a lipid-raft-dependent signaling complex in human monocytes." *Blood (106:3074–3081)*:DOI. See pp. 108, 110, 128, 129.

SOP

1. SOP PBMC

SOP

SOP_Patient_PBMC_and_Plasma_collecting.rtf

Patient PBMC & plasma collecting procedure

A) General notes

- The whole procedure needs to be conducted under sterile conditions.
- For all the centrifugation steps use only centrifuges with swinging rotors and with acceleration / brake adjustment possibility to avoid sample loss.
- Carry out the complete procedure at room temperature (RT).
- Use only PBS and FICOLL at RT.
- Samples are to be labeled with a permanent marker as follows (example for the first patient):

Plasma (P)   Cells (Z)

LU#0001   LU#0001
Date   Date
P   Z

Use progressive patient numbering and keep patient data record.

B) Material preparation

1. Prepare and label (see above) per patient: 6 cryo tubes (4 for cells & 2 for plasma), 2 * 50ml Falcons, and 4 * 5ml Falcons.
2. Take FICOLL stock out of the fridge and pipette 15ml of FICOLL to one 50ml Falcon tube and allow FICOLL to reach RT before proceeding.
3. Make sure to have enough RT PBS (~80ml per patient).

C) Plasma sampling

4. Centrifuge the 4 tubes of patient blood 5min @ 770g, RT with brake and normal acceleration.
5. Mark each tube at the surface level of the sample to register the total volume.
6. Carefully aspirate plasma (yellow layer) of 4 patient blood tubes and transfer to 4 * 5ml Falcon tubes – do not aspirate completely to avoid taking white blood cells at the blood-plasma-interface.

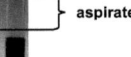
aspirate

7. Take 0.5ml of each of the 4 * 5ml Falcon tubes and transfer to 2 properly labeled ("P") cryo tubes (final 1ml each).

D) Cell sampling

8. Give back taken off plasma volume in RT PBS to the total volume mark and mix gently by pipetting.
9. **CAREFULLY** layer all 4 RT PBS reconstituted patient blood samples (one by one) on top of the FICOLL (in one 50ml Falcon tube).
10. Centrifuge the 50ml FICOLL tube of patient blood 20min @ 770g, RT **WITH SLOW ACCELERATION** and **WITHOUT BRAKE**.
11. Collect ring of white blood cells (WBZ) to 1 new 50ml Falcon **WITHOUT** touching the red blood cells (RBZ).

02.03.2007/mfc

SOP PBMC

SOP_Patient_PBMC_and_Plasma_collecting.rtf

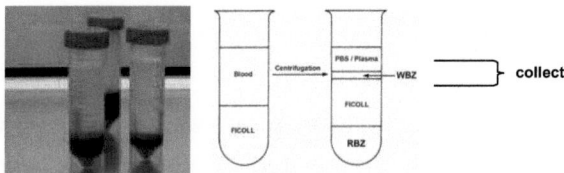

12. Fill 50ml Falcon tube ad 50ml with RT PBS.
13. Centrifuge the 50ml Falcon tube 5min @ 770g, RT with normal acceleration and brake.
14. A cell pellet MUST be visible.
15. Discard supernatant (SN) by pouring (carefully).
16. Resuspend pellet in 4ml 4°C freezing medium (just taken out of the fridge before usage).
17. Aliquot 4 * 1ml of resuspension to 4 properly labeled ("Z") cryo tubes.

E) *Plasma & Cell freezing*

18. Place all 6 cryo tubes into **styropor box**, close box well, and freeze @ -70°C in the back of the freezer together with the plasma samples (5ml Falcons).
19. The samples should remain in the freezer for at least 2 days before shipping and can be stored up to 4 weeks @ -70°C.
20. Organize shipment on dry ice after collection of several patient samples together with the Experimental Immunology lab (DF USB, Tel. 061 265 23 27, Fax 061 265 23 50, marco.cavallari@unibas.ch, lucia.mori@unibas.ch) and the Signal Transduction lab (DF USB, Tel. 061 265 24 22, Fax 061 265 23 50, Therese-J.Resink@unibas.ch).

F) *Material references*

- Falcon tubes (BDbiosciences.com, 352070 (50ml) & 352063 (5ml))

- Cryo tubes (Nuncbrand.com, 375418 (1.8ml))

- FICOLL (Axon Lab, Lymphoprep) to be stored @ 4°C in the dark (alufoil wrapped)
- PBS (Invitrogen.com (Gibco), 14190)
- Freezing medium: 90% fetal bovine serum (FBS) & 10% DMSO to be stored @ 4°C

G) *General references*

55% plasma
5% white blood cells   } Blood components
40% red blood cells & platelets

02.03.2007/mfc

197

H) Protocol at a glance

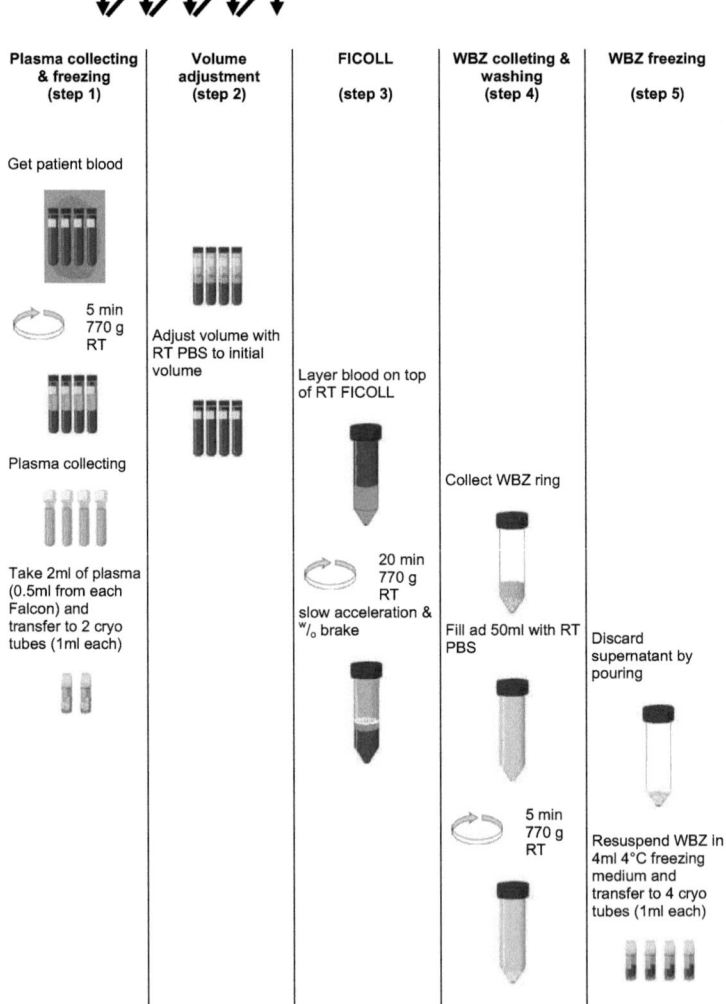

2. SOP mycoplasma

SOP

mycoplasma_detection_20070227.doc

Mycoplasma detection procedure DF, USB

I) Purpose

The purpose of this Standard Operating Procedure (SOP) is to establish uniform procedures for ensuring that
- Newly arriving cells or cell lines are screened upon arrival (not compulsory for commercially purchased cells or cell lines that are routinely tested and certified mycoplasma-free by the supplier, e.g. ATCC, ECACC, DSMZ, …).
- All cell culture facilities of the DF are routinely checked for mycoplasma contaminations.
- All cell culture facilities of the DF (become and) remain mycoplasma-free.
- All liquid nitrogen tanks of the DF (become and) remain mycoplasma-free.
- …

II) Applicability & Compliance

The procedure contained in, or referenced by, this SOP is applicable to all personnel in the DF cell culture facilities.
Compliance evaluation inspections at the DF will be conducted in a manner which will maintain the existing level of bio-safety at the facility being inspected.

III) Summary of Procedure

The mycoplasma screening and possible decontamination will be accomplished in three stages
- Pre-screening cell (and potential mycoplasma) expansion and DNA isolation
- Screening by genomic PCR
- Post-screening (if mycoplasma-positive) attempt to eliminate the contamination(s)

IV) Definitions

The following definitions are applicable to this SOP.
- *Mycoplasma*: Concerning a first reference to mycoplasma it is suggested to read Appendix A (taken from http://en.wikipedia.org/wiki/Mycoplasma 27th of February 2007).
- *Bio-safety & GLP*: The practices and procedures (in accordance with the "Organization and Guidelines of the DF") which if implemented are likely to prevent or greatly reduce the likelihood of the spread of a mycoplasma contamination.
- *Health and Safety Warnings*: As dictated by the "Organization and Guidelines of the DF" and as laid out in the operational manuals specific to the equipment in use.

15.03.2007/mfc

V) The adapted SOP (Standard Operating Procedure) for the DF

a) Pre-screening cell (and potential mycoplasma) expansion and DNA isolation

All arriving cells or cell lines have to be tested upon arrival (not compulsory for commercially purchased cells or cell lines that are routinely tested and certified mycoplasma-free by the supplier, e.g. ATCC, ECACC, DSMZ, …).
For mycoplasma screening culture cells (if applicable in BSL2 incubator) for **more than 1 week** in your medium of choice.

Extract DNA from $1\text{-}2*10^6$ cells (in culture > 1 week after thawing) with NucleoSpin Blood acc. to manufacturers protocol (see Appendix B for the "Protocol at a glance rev04" page 2).
Store eluted DNA acc. to NucleoSpin Blood recommendations (e.g. -20°C).

Prepare and include **all proper PCR controls** to assure DNA integrity and PCR functionality, so you will have a minimum set of 4 PCR reactions to test 1 cell line as follows:
1. **DNA control**: test DNA plus a set of genomic DNA control primers
2. **mycoplasma positive control**: mycoplasma positive DNA sample plus mycoplasma primers
3. **mycoplasma negative control**: mycoplasma negative DNA sample plus mycoplasma primers
4. **mycoplasma test**: test DNA plus mycoplasma primers

b) Screening by genomic PCR

Reaction mix
10x ThermoPol buffer (NEB, B9004S)   2 µl
10x (2mM) dNTP   2 µl
Primer 10µM each   1 + 1 = 2 µl
H$_2$O   12.8 µl
Template   1 µl
TaqPol (NEB, M0267L) 5000U/ml   0.2 µl   (see Appendix C for TDS)
Total volume   20µl

Mycoplasma PCR
95°C   5'
   95°C   40" ⎫
   60°C   40" ⎬   30 ↻
   72°C   40" ⎭
72°C   5'
4°C   pause

Primers tested
5' Myco_9   5'   CGC CTG AGT AGT ACG TTC GC   3'
3' Myco_3   5'   GCG GTG TGT ACA AGA CCC GA   3'
product size: 500bp

Primers not yet tested
Primer fwd   5'   ACA CCA TGG GAG CTG GTA AT   3'
Primer rev   5'   CTT CAT CGA CTT TCA GAC CCA AGG CAT   3'
product size: ~300 – 400bp

KB013 Myco_F1   5'   ACA CCA TGG GAG CTG GTA AT   3'
KB014 Myco_R1   5'   CTT CAT CGA CTT CCA GAC CCA AGG CAT   3'
product size: ~500bp (16S-23S rRNA spacer)

Run gel (usually 1% agarose) of PCR reaction(s).

15.03.2007/mfc

c) Post-screening (if mycoplasma-positive) attempt to eliminate the contamination(s)

For mycoplasma contaminated cells (if applicable in BSL2 incubator) try to cure with 25 µg/ml Plasmocin for **2 weeks** in your medium of choice (see Appendix D for manufacturer's protocol and TDS).
Retest cells according to the SOP directly after Plasmocin treatment and another time more than 2 weeks after the end of the treatment to assure the cells remain mycoplasma-free.
Use and freeze cells only after second mycoplasma-negative test.

d) Concluding remarks and further suggestions

Do not forget to clean up the liquid nitrogen from mycoplasma-positive freezing batches.
Use Plasmocure (Invivogen) instead of Plasmocin (Invivogen).

VI) References

The reference papers to the mycoplasma detection method used in the Experimental Immunolgy Lab of the DF are
- Hopert A et al.: J Immunol Methods. 1993 Aug 26;164(1):91-100.
- Hopert A et al.: In Vitro Cell Dev Biol Anim. 1993 Oct;29A(10):819-21.

mycoplasma_detection_20070227.doc

VII) Appendices
a) Mycoplasma_wiki_20070227.pdf

The reference was taken from http://en.wikipedia.org/wiki/Mycoplasma the 27th of February 2007 (note that only part of the wikipedia entry is printed below).

Mycoplasma is a genus of bacteria that lack cell walls. They can be parasitic or saprophytic. Several species are pathogenic in humans, including *M. pneumoniae*, which is an important cause of pneumonia and other respiratory disorders, and *M. genitalium*, which is believed to be involved in pelvic inflammatory diseases. They are unaffected by antibiotics that target cell wall synthesis, such as penicillin.

The genus *Mycoplasma* is one of several genera within the class *Mollicutes*. Mollicutes are bacteria which have small genomes, lack a cell wall and have low GC-content (18-40 mol%). There are over 100 recognized species of the genus *Mycoplasma*. Their genome size ranges from 0.6 - 1.35 megabase-pairs. Mollicutes are parasites or commensals of humans, other animals including insects, and plants; the genus *Mycoplasma* is by definition restricted to vertebrate hosts. Cholesterol is required for the growth of species of the genus *Mycoplasma* as well as certain other genera of mollicutes. Their optimum growth temperature is often the temperature of their host if warmbodied (e.g. 37 degrees Celsius in humans) or ambient temperature if the host is unable to regulate its own internal temperature. Analysis of 16S ribosomal RNA sequences as well as gene content strongly suggest that the mollicutes, including the mycoplasmas, are closely related to either the *Lactobacillus* or the *Clostridium* branch of the phylogenetic tree (Firmicutes *sensu stricto*).

Mycoplasmas are often found in research laboratories as contaminants in cell culture and come from careless handling; due to their small size, they are difficult to detect under a microscope and their presence can skew experimental results.

15.03.2007/mfc

SOP

mycoplasma_detection_20070227.doc

b) *NucleoSpin Blood (MN) GenomicDNABlood_R04 p2.pdf*

Protocol at a glance (Rev. 04)
Genomic DNA Purification from Blood

		Mini — NucleoSpin® Blood	Midi — NucleoSpin® Blood L	Maxi — NucleoSpin® Blood XL	Mini — NucleoSpin® Blood QuickPure
1	Lyse blood samples	200 µl blood 25 µl prot. K 200 µl B3 Mix 70°C 10-15 min	2 ml blood 150 µl prot. K 2 ml BQ1 Mix 56°C 15 min	10 ml blood 500 µl prot. K 10 ml BQ1 Mix 56°C 15 min	200 µl blood 25 µl prot. K 200 µl BQ1 Mix 70°C 10-15 min
2	Adjust DNA binding conditions	210 µl ethanol	2 ml ethanol	10 ml ethanol	200 µl ethanol
3	Bind DNA	load all 1 min 11,000 x g	load 3 ml 3 min 4,500 x g load 3 ml of residue 5 min 4,500 x g	load 15 ml 3 min 4,000 x g load 15 ml of residue 3 min 4,000 x g	load all 1 min 11,000 x g
4	Wash silica membrane	500 µl BW 600 µl B5 1st wash: 1 min 11,000 x g 2nd wash: 1 min 11,000 x g	2 ml BQ2 2 ml BQ2 1st wash: 2 min 4,500 x g 2nd wash: 10 min 4,500 x g	7.5 ml BQ2 7.5 ml BQ2 1st wash: 2 min 4,000 x g 2nd wash: 10 min 4,000 x g	350 µl BQ2 1st wash: 3 min 11,000 x g
5	Dry silica membrane	1 min 11,000 x g	Drying is performed during centrifugation of the last washing step	Drying is performed during centrifugation of the last washing step	Drying is performed during centrifugation of the single washing step
6	Elute highly pure DNA	100 µl BE (70°C) RT 1 min 1 min 11,000 x g	200 µl BE (70°C) RT 2 min 2 min 4,500 x g	500-2000 µl BE (70°C) RT 2 min 2 min 4,000 x g	50 µl BE (70°C) RT 1 min 1 min 11,000 x g

15.03.2007/mfc

SOP mycoplasma

mycoplasma_detection_20070227.doc

c) Taq M0267_NEB.pdf

New England Biolabs, Inc.
Tel: 800-632-5227 (orders)
Tel: 800-632-7799 (support)
Fax: 978-921-1350
e-mail: info@neb.com
WWW: http://www.neb.com

Taq DNA Polymerase with ThermoPol Buffer

#M0267S   400 units .... $50 (USA)
#M0267L   2,000 units ...... $200 (USA)

Description
Taq DNA Polymerase is a thermostable DNA polymerase that possesses a non-processive 5´→ 3´ polymerase activity and a double-strand specific 5´→ 3´ exonuclease activity.

It is supplied with 10X ThermoPol Reaction Buffer, which contains a nonionic detergent to increase enzyme stability during longer incubations.

Source
An *E. coli* strain that carries the *Taq* DNA Polymerase gene from the *Thermus aquaticus* YT-1.

Application
- PCR
- Primer Extension
- Colony PCR
- Long PCR (> 5 kb)

Reaction Conditions:
1X ThermoPol Reaction Buffer, DNA template, primer, 200 µM dNTPs and 2–5 units of *Taq* DNA Polymerase in a total reaction volume of 100 µl.

Reaction Buffer
1X ThermoPol Reaction Buffer:
[20 mM Tris-HCl (pH 8.8, @ 25°C), 10 mM KCl, 10 mM $(NH_4)_2SO_4$, 2 mM $MgSO_4$, 0.1% Triton X-100].
Supplied with enzyme as a 10X concentrated stock.

Concentration
5,000 units/ml

Storage Conditions
100 mM KCl, 10 mM Tris-HCl (pH 7.4), 1 mM EDTA, 1 mM dithiothreitol, 0.5% Tween 20, NP-40 and 50% glycerol.
Store at –20°C.

Quality Assurance
Purified free of contaminating endonucleases and exonucleases.

Unit Definition
One unit is defined as the amount of enzyme that will incorporate 10 nmol of dNTP into acid-insoluble material in 30 minutes at 75°C.

Unit Assay Conditions
1X ThermoPol Reaction Buffer, 200 µM each dNTP including [^{3}H]-dTTP, 200 µg/ml activated calf thymus DNA.

Patents/Disclaimer:
Some applications in which this product can be used may be covered by patents issued and applicable in the United States and certain other countries. Because purchase of this product does not include a license to perform any patented application, users of this product may be required to obtain a patent license depending upon the particular application in which the product is used. The PCR process is the subject of European Patent Nos. 201,184 and 2000,262 owned by Hoffman-LaRoche. Those patents will expire on March 28, 2006. The corresponding PCR process patents in the United States expired on March 29, 2005.

Technical Bulletin
#M0267(4/11/05)

15.03.2007/mfc

SOP

mycoplasma_detection_20070227.doc

d) Plasmocin (Invivogen)_TDS.pdf

Plasmocin™

For treatment and prevention of Mycoplasma contamination of cell culture
Catalog # ant-mpt, ant-mpp

For research use only
Version # 02F12-EL

PRODUCT INFORMATION
Content:
Plasmocin™ is supplied as 1 ml tubes containing a sterile, yellow solution, cell culture tested.
- **ant-mpt:** 2x 1ml at 25mg/ml (50mg). For elimination of Mycoplasma contaminants.
- **ant-mpp:** 5x 2ml at 2.5mg/ml (25mg). For prevention of Mycoplasma contamination.

Shipping and Storage:
Plasmocin™ is shipped at room temperature. Upon receipt, it should be stored at 4°C for immediate use and is stable at this temperature for several weeks. If Plasmocin™ is not to be used immediately, then freeze at -20°C for long term storage.
Plasmocin™ is stable for at least one year when stored at -20°C.
Note: Presence of crystals does not alter properties of the product. Vortex the tube until the crystals disappear.

Quality Control:
Activity of Plasmocin™ is rigorously controlled by physicochemical and microbiological assays.

GENERAL PRODUCT USE
Plasmocin™ is used to cure cell lines infected by Mycoplasma and related cell wall-less bacteria. Plasmocin™ can also be used as a routine addition in liquid media to prevent Mycoplasma and more generally bacterial contamination in small and large animal cell cultures.

BACKGROUND
Recent reports estimate Mycoplasma contamination in up to 63% of all cell cultures. Mycoplasma cannot be detected by visual inspection and may not noticeably affect cell culture growth rates. However, Mycoplasma infection has been shown to alter DNA, RNA and protein synthesis, introduce chromosomal aberrations and cause alterations or modifications of host cell plasma membrane antigens.
Plasmocin™ is a new generation of bactericidal antibiotic preparation strongly active on Mycoplasma infected cells. It is active at low concentrations on a broad range of gram positive and gram negative bacteria otherwise resistant to the mixture of streptomycin and penicillin antibiotics commonly used in cell cultures.

DESCRIPTION / PROPERTIES
Plasmocin™ contains two newly developed bactericidal components: one acts on the protein synthesis machinery by interfering with ribosome translation, and the other acts on DNA replication by interfering with the replication fork. These two specific and separate targets are found only in Mycoplasma and many other bacteria, and are completely absent in eukaryotic cells.

In contrast to most anti-Mycoplasma compounds that act solely *in vitro*, Plasmocin™ is active on Mycoplasma present in cell culture medium, and on intracellular Mycoplasma found in some specialized mammalian cells. The two antibiotics comprising Plasmocin™ are actively transported into mammalian cells providing a synergistic killing effect on intracellular Mycoplasma without any apparent adverse effect on cellular metabolism. This benefit insures that after being treated with Plasmocin™, a cell culture is not reinfected by Mycoplasma released from the intracellular compartments of infected cells following antibiotic removal. At high concentrations of Plasmocin™, slowdown of cell growth rate may be observed. This slowing down is mainly due to the inhibition of mitochondria respiration by Plasmocin™. However when Plasmocin™ is removed from culture medium, cells return rapidly to their normal growth rate. The anti-Mycoplasma activity of Plasmocin™ is unaltered in cell culture medium containing up to 20% serum.

RESISTANCE TO PLASMOCIN™
In repeated experiments aimed to determine the mutation rate of *Mycoplasma hominis*, *Mycoplasma bovis* and *Acholeplasma vituli* to Plasmocin™, no resistance in liquid cultures has ever been identified, indicating a possible mutation rate lower than 10^{-9}. Therefore, development of resistant Mycoplasma strains is highly unlikely.

METHOD
Treatment of Mycoplasma Infected Cell Cultures:
Plasmocin™ treatment (ant-mpt) requires little hands-on manipulation and is completed in only two weeks. Typically, Plasmocin™ is used at 25 µg/ml which represents a 1:1000 dilution of the 25 mg/ml stock solution. Working concentration of Plasmocin™ ranges from 12.5 to 37.5 µg/ml.

1- Split an actively dividing culture of cells into medium containing 12-25µg/ml of Plasmocin™.
2- Remove and replace with fresh Plasmocin™ containing medium every 3-4 days for 2 weeks.
3. For maintenance of a Mycoplasma free culture, continue the use of Plasmocin™ at a concentration of 5µg/ml.
Note: If Mycoplasma elimination is not completed after a two week treatment, you may continue the treatment for an additional week and/or increase the concentration to 37.5 µg/ml.

Maintenance or prophylactic use against Mycoplasma infections:
To prevent Mycoplasma and related cell wall-less bacteria contaminations of cell cultures that have been previously tested to be contamination-free, use Plasmocin™ prevention (ant-mpp) at a concentration of 5 µg/ml that represents a 1:500 dilution of the 2.5 mg/ml stock solution.

TECHNICAL SUPPORT
Toll free (US): 888-457-5873
Outside US: (+1) 858-457-5873
E-mail: info@invivogen.com
Website: www.invivogen.com

3950 Sorrento Valley Blvd. Suite A
San Diego, CA 92121 - USA

15.03.2007/mfc

Documents

3. Creative Commons License

Creative Commons

Creative Commons License Deed

Attribution-NoDerivs 2.5 Switzerland (CC BY-ND 2.5)

This is a human-readable summary of the Legal Code (the full license) available in the following languages:
German
Disclaimer

You are free:

to Share — to copy, distribute and transmit the work

to make commercial use of the work

Under the following conditions:

Attribution — You must attribute the work in the manner specified by the author or licensor (but not in any way that suggests that they endorse you or your use of the work).

No Derivative Works — You may not alter, transform, or build upon this work.

With the understanding that:

Waiver — Any of the above conditions can be waived if you get permission from the copyright holder.

Public Domain — Where the work or any of its elements is in the public domain under applicable law, that status is in no way affected by the license.

Other Rights — In no way are any of the following rights affected by the license:
- Your fair dealing or fair use rights, or other applicable copyright exceptions and limitations;
- The author's moral rights;
- Rights other persons may have either in the work itself or in how the work is used, such as publicity or privacy rights.

- **Notice** — For any reuse or distribution, you must make clear to others the license terms of this work. The best way to do this is with a link to this web page.

i want morebooks!

Buy your books fast and straightforward online - at one of world's fastest growing online book stores! Environmentally sound due to Print-on-Demand technologies.

Buy your books online at
www.get-morebooks.com

Kaufen Sie Ihre Bücher schnell und unkompliziert online – auf einer der am schnellsten wachsenden Buchhandelsplattformen weltweit! Dank Print-On-Demand umwelt- und ressourcenschonend produziert.

Bücher schneller online kaufen
www.morebooks.de

VDM Verlagsservicegesellschaft mbH
Heinrich-Böcking-Str. 6-8　　Telefon: +49 681 3720 174　　info@vdm-vsg.de
D - 66121 Saarbrücken　　　 Telefax: +49 681 3720 1749　　www.vdm-vsg.de

Printed by Books on Demand GmbH, Norderstedt / Germany